Ma

Practice in the Basic Skills

Contents

Addition

A	40	46	32	26	33	84	53
	26	11	26	51	42	13	41
	+12	+22	+31	+22	+14	+ 1	+ 4
B	264	251	871	524	749	410	555
	103	46	14	355	120	368	203
	+622	+502	+104	+120	+ 30	+210	+121
C	45	32	85	49	24	63	52
	27	28	9	18	49	26	17
	+26	+27	+15	+13	+29	+ 7	+19
D	267	423	504	387	423	222	709
	108	347	264	207	67	548	159
	+519	+216	+225	+204	+408	+ 26	+131
E	62	83	94	63	92	75	73
	43	56	24	64	86	62	22
	+71	+60	+71	+81	+41	+42	+74
F	242	302	621	294	781	542	270
	70	482	193	182	192	333	477
	+635	+172	+154	+292	+ 6	+ 84	+ 81
G	45	56	78	29	63	39	86
	37	48	20	87	49	17	70
	+49	+39	+57	+38	+25	+89	+37
H	429	362	96	527	492	187	209
	76	249	287	209	271	483	385
	+147	+158	+402	+186	+ 39	+186	+294

Addition

A	331	127	236	624	842	705
	840	931	543	601	724	632
	+526	+341	+820	+253	+413	+252
B	247	538	867	328	835	424
	726	227	278	305	347	657
	+485	+993	+654	+770	+562	+206
C	909	722	474	200	742	526
	236	337	826	439	758	437
	+474	+560	+658	+527	+214	+485
D	862	925	720	536	453	304
	604	859	673	950	822	957
	+973	+980	+858	+707	+968	+983
E	425	739	729	651	945	472
	808	716	937	614	867	870
	+802	+528	+462	+939	+445	+763
F	510	954	865	697	896	482
	827	927	794	485	813	958
	+735	+483	+978	+891	+824	+761

G	1431	2627	5003	9040	3214
	2627	1393	2435	435	429
	+ 431	+ 729	+1620	+ 274	+4005
H	627	1109	2408	5406	643
	5213	4329	770	603	7554
	+2471	+3491	+6378	+2654	+1500

Subtraction

A	963 –421	899 –382	764 –241	869 – 53	427 –306	588 –337
B	974 –673	659 – 58	432 –331	784 –482	904 –503	747 –346
C	843 –643	978 –678	503 –403	726 –326	298 –198	629 –229
D	530 –217	780 –358	920 –416	860 –345	680 –257	450 –139
E	607 –436	908 –564	306 –273	505 –354	403 –291	804 –684
F	436 –227	924 –508	691 –479	885 –466	543 –225	742 –334
G	539 –346	428 –275	917 –664	646 –152	354 –161	866 –593
H	273 –178	964 –676	738 –459	453 –288	647 –279	385 –287
I	300 –173	700 –486	600 –592	800 –767	500 –289	900 –859
J	607 –488	408 –139	202 – 43	805 –567	706 –528	903 –745

K Complete the number sentences.

15 – ☐ = 7	32 – 7 = ☐	32 + ☐ = 41	☐ – 9 = 16
22 – ☐ = 6	46 – 9 = ☐	16 + ☐ = 23	☐ – 8 = 15
17 – ☐ = 9	51 – 7 = ☐	18 + ☐ = 31	☐ – 7 = 13

Subtraction

A	1047 – 523	1036 – 412	1029 – 908	1065 – 822	1054 – 741	1018 – 307
B	1276 – 540	1583 – 670	1469 – 825	1798 – 974	1657 – 710	1145 – 421
C	1859 – 976	1218 – 327	1736 – 853	1327 – 552	1145 – 871	1463 – 792
D	1157 – 258	1861 – 974	1618 – 949	1532 – 756	1423 – 648	1344 – 888
E	1008 – 359	1005 – 476	1004 – 528	1003 – 875	1002 – 736	1006 – 437
F	1000 – 435	1000 – 267	1000 – 875	1000 – 492	1000 – 814	1000 – 458
G	4027 –2471	7045 –5427	8089 –3947	2038 –1432	3067 –1745	6089 –2796
H	5386 –4827	4221 –2987	5167 –1496	9435 –3587	7549 –4592	9620 –4892
I	4000 –2532	7000 –3726	6000 –1847	3000 –2654	8000 –4359	5000 –4629
J	6005 –4739	4007 –2518	8006 –2929	3002 –1627	9005 –3426	7002 –6109

K Write answers only.

62 minus 15	52 subtract 9	24 take 6	56 less 27
43 minus 8	27 subtract 18	61 take 34	23 less 14
32 minus 7	22 subtract 7	44 take 16	76 less 39

Notation

A Write the value of the figure underlined in each number.

$\underline{8}027$	$4\underline{3}5$	$3\underline{4}27$	$7\underline{6}$	$4\underline{2}32$
$65\underline{2}3$	$\underline{4}231$	$941\underline{6}$	$\underline{5}69$	$34\underline{1}0$

B Write in figures:

six hundred and twenty	sixty
eight hundred and fifty	eighty-five
one hundred and fifteen	ninety
seven hundred and ten	two hundred
six thousand and fifty-four	forty-nine
three thousand and six	nine hundred and six

C Write in words:

4007	793	115	505	9620	20
30	402	530	107	3017	999

D

Add 1 to:	99	409	420	1199
Take 1 from:	120	143	60	1230
Add 10 to:	193	2290	186	993
Take 10 from:	103	174	3077	706
Add 100 to:	941	2304	1721	274
Take 100 from:	572	2341	4831	2000

Addition and subtraction

A Write in columns then add.

47 + 276 + 3471

1347 + 23 + 547

2163 + 35 + 7

216 + 359 + 42 + 3129

4132 + 3671 + 26 + 9

8021 + 1021 + 104 + 15

741 + 603 + 1432 + 43

2916 + 356 + 7

2135 + 4236 + 17

4039 + 376 + 56

B Write in columns then subtract.

4271 – 369

1402 – 347

8372 – 7

707 – 45

3271 – 2937

4021 – 376

739 – 27

496 – 171

2036 – 19

3241 – 39

5436 – 207

6271 – 499

C Find the sum of 46, 325, 4216.

Find the total of 3216, 472, 27 and 9.

96 plus 376 plus 4298 plus 6.

To the sum of 46 and 217 add 767.

To 7421 add the sum of 96 and 432.

D How much less than 2063 is 1479?

By how much is 734 greater than 529?

By how many is 76 less than 1006?

4326 minus 1023.

From 476 subtract 98.

Multiplication

A	424 ×2	112 ×3	122 ×4	333 ×3	322 ×3	212 ×4
B	102 ×5	107 ×6	208 ×4	309 ×2	206 ×3	106 ×6
C	227 ×3	346 ×2	119 ×5	115 ×6	117 ×5	124 ×4
D	390 ×2	151 ×6	172 ×4	191 ×5	182 ×4	283 ×3
E	97 ×6	234 ×3	498 ×2	189 ×5	219 ×4	89 ×6
F	425 ×6	370 ×5	293 ×4	746 ×3	598 ×2	649 ×6
G	1471 ×5	2362 ×3	1625 ×4	1063 ×6	1549 ×2	

H	1361 × 4	1403 × 5	2347 × 2	1940 × 3
	796 × 6	1079 × 5	3499 × 2	999 × 6

I Multiply 276 by 5.
Multiply 1274 by 4.
Multiply 396 by 6.
Find the product of 76 and 5.
Find the product of 143 and 4.
Find the product of 98 and 6.

J What number is three times two hundred and five?
Multiply three hundred and fifty by four.
Multiply together six and ninety-six.

Multiplication by 7

A

$3 \times 7 = \square$	$7 \times 2 = \square$	$4 \times 7 = \square$	$7 \times 6 = \square$	$9 \times 7 = \square$
$12 \times 7 = \square$	$5 \times 7 = \square$	$8 \times 7 = \square$	$7 \times 3 = \square$	$2 \times 7 = \square$
$1 \times 7 = \square$	$10 \times 7 = \square$	$7 \times 7 = \square$	$6 \times 7 = \square$	$7 \times 5 = \square$
$0 \times 7 = \square$	$7 \times 4 = \square$	$11 \times 7 = \square$		

B

$(3 \times 7) + 6 = \square$	$(2 \times 7) + 5 = \square$	$(4 \times 7) + 3 = \square$	$(5 \times 7) + 4 = \square$
$(9 \times 7) + 2 = \square$	$(7 \times 7) + 5 = \square$	$(6 \times 7) + 2 = \square$	$(8 \times 7) + 5 = \square$
$(3 \times 7) + 3 = \square$	$(6 \times 7) + 3 = \square$	$(8 \times 7) + 4 = \square$	$(2 \times 7) + 6 = \square$
$(4 \times 7) + 5 = \square$	$(7 \times 7) + 4 = \square$	$(5 \times 7) + 2 = \square$	$(9 \times 7) + 6 = \square$
$(3 \times 7) + 4 = \square$	$(8 \times 7) + 6 = \square$	$(9 \times 7) + 4 = \square$	$(4 \times 7) + 6 = \square$
$(2 \times 7) + 2 = \square$	$(5 \times 7) + 3 = \square$	$(7 \times 7) + 3 = \square$	

C

426 ×7	396 ×7	457 ×7	187 ×7	609 ×7	148 ×7
324 ×7	635 ×7	298 ×7	439 ×7	177 ×7	208 ×7
1235 ×7	1389 ×7	1176 ×7	1068 ×7	1403 ×7	

D

473×7	629×7	808×7	929×7
1143×7	1086×7	1205×7	1097×7

E Find the product of 497 and 7.

Multiply five hundred and nine by seven.

What number is seven times eighty-seven?

Find the product of 379 and 7.

What number is seven times ninety-six?

Multiply three hundred and ten by seven.

Multiplication by 8

A

$\square \times 8 = 16$	$\square \times 8 = 64$	$\square \times 8 = 24$	$\square \times 8 = 48$	$\square \times 8 = 96$
$\square \times 8 = 0$	$8 \times \square = 32$	$\square \times 8 = 56$	$\square \times 8 = 40$	$\square \times 8 = 88$
$\square \times 1 = 8$	$8 \times \square = 24$	$8 \times \square = 48$	$\square \times 8 = 32$	$\square \times 8 = 80$
$8 \times \square = 16$	$\square \times 8 = 72$	$8 \times \square = 8$	$8 \times \square = 56$	$8 \times \square = 40$

B Add:

$64 + 6 = \square$	$48 + 2 = \square$	$16 + 7 = \square$	$24 + 6 = \square$	$8 + 2 = \square$
$56 + 7 = \square$	$48 + 3 = \square$	$8 + 4 = \square$	$56 + 6 = \square$	$64 + 7 = \square$
$48 + 5 = \square$	$8 + 6 = \square$	$56 + 4 = \square$	$24 + 7 = \square$	$48 + 4 = \square$
$16 + 5 = \square$	$8 + 7 = \square$	$16 + 6 = \square$	$48 + 6 = \square$	$8 + 5 = \square$
$16 + 4 = \square$	$48 + 7 = \square$	$56 + 5 = \square$	$8 + 3 = \square$	

C

327 ×8	409 ×8	618 ×8	520 ×8	987 ×8	431 ×8
606 ×8	897 ×8	532 ×8	170 ×8	547 ×8	823 ×8
1098 ×8	1209 ×8	1070 ×8	1165 ×8	1143 ×8	

D

380×8	675×8	989×8	206×8
1149×8	1069×8	1006×8	1124×8

E Find the product of: 397 and 8 1027 and 8 729 and 8

Multiply by eight:

439 1025 929 630 836 1009 606

Multiplication by 9

A

$1 \times 9 = \square$ $9 \times 4 = \square$ $6 \times 9 = \square$ $7 \times 9 = \square$ $0 \times 9 = \square$

$9 \times 5 = \square$ $9 \times 8 = \square$ $4 \times 9 = \square$ $12 \times 9 = \square$ $9 \times 9 = \square$

$9 \times 7 = \square$ $3 \times 9 = \square$ $11 \times 9 = \square$ $2 \times 9 = \square$ $9 \times 2 = \square$

$9 \times 0 = \square$ $5 \times 9 = \square$ $8 \times 9 = \square$ $10 \times 9 = \square$ $9 \times 3 = \square$

$9 \times 6 = \square$

B Add:

$9 + 2 = \square$ $36 + 7 = \square$ $36 + 6 = \square$ $27 + 4 = \square$ $31 + 9 = \square$

$9 + 6 = \square$ $63 + 8 = \square$ $63 + 7 = \square$ $27 + 8 = \square$ $72 + 9 = \square$

$27 + 9 = \square$ $18 + 7 = \square$ $27 + 6 = \square$ $9 + 3 = \square$ $54 + 7 = \square$

$36 + 8 = \square$ $27 + 7 = \square$ $45 + 9 = \square$ $72 + 8 = \square$ $27 + 5 = \square$

$9 + 5 = \square$ $18 + 6 = \square$ $9 + 9 = \square$ $18 + 5 = \square$ $36 + 5 = \square$

$45 + 5 = \square$ $63 + 9 = \square$ $18 + 4 = \square$ $9 + 8 = \square$ $45 + 6 = \square$

$45 + 8 = \square$ $54 + 8 = \square$ $18 + 9 = \square$ $54 + 6 = \square$ $54 + 9 = \square$

$18 + 8 = \square$ $9 + 7 = \square$ $36 + 4 = \square$ $18 + 2 = \square$ $45 + 7 = \square$

$27 + 3 = \square$ $9 + 4 = \square$ $36 + 9 = \square$ $9 + 1 = \square$ $18 + 3 = \square$

C

427	683	939	157	708	526
×9	×9	×9	×9	×9	×9

631	760	909	386	825	431
×9	×9	×9	×9	×9	×9

1076	1107	1035	1108
×9	×9	×9	×9

D Copy and fill in the missing numbers.

4 ☐	8 ☐ 9	2 5 3	1 ☐ 4 ☐
×9	×9	×9	×9
☐ 1 4	7 2 8 1	2 ☐ 7 ☐	☐ 4 0 5

Multiplication by 11

A

$7 \times 11 = \square$	$4 \times 11 = \square$	$3 \times 11 = \square$	$1 \times 11 = \square$	$9 \times 11 = \square$
$8 \times 11 = \square$	$2 \times 11 = \square$	$6 \times 11 = \square$	$5 \times 11 = \square$	$0 \times 11 = \square$
$11 \times \square = 88$	$11 \times \square = 44$	$11 \times \square = 55$	$11 \times \square = 11$	$11 \times \square = 66$
$11 \times \square = 33$	$11 \times \square = 77$	$11 \times \square = 0$	$11 \times \square = 99$	$11 \times \square = 22$

B Add:

$77 + 4 = \square$	$88 + 8 = \square$	$44 + 7 = \square$	$77 + 5 = \square$	$66 + 6 = \square$
$77 + 3 = \square$	$33 + 9 = \square$	$77 + 6 = \square$	$88 + 9 = \square$	$55 + 9 = \square$
$66 + 7 = \square$	$88 + 6 = \square$	$77 + 7 = \square$	$55 + 6 = \square$	$99 + 9 = \square$
$88 + 7 = \square$	$99 + 8 = \square$	$33 + 7 = \square$	$66 + 9 = \square$	$22 + 9 = \square$
$55 + 5 = \square$	$22 + 8 = \square$	$77 + 9 = \square$	$66 + 8 = \square$	$55 + 7 = \square$
$99 + 7 = \square$	$55 + 8 = \square$	$66 + 5 = \square$	$99 + 5 = \square$	$88 + 3 = \square$
$33 + 8 = \square$	$77 + 8 = \square$	$44 + 9 = \square$	$99 + 6 = \square$	$44 + 8 = \square$
$99 + 2 = \square$	$88 + 4 = \square$	$99 + 3 = \square$	$88 + 5 = \square$	$88 + 2 = \square$
$99 + 1 = \square$	$11 + 9 = \square$	$66 + 4 = \square$	$44 + 6 = \square$	$99 + 4 = \square$

C

235	404	678	172	287	356
×11	×11	×11	×11	×11	×11
140	906	825	732	624	382
×11	×11	×11	×11	×11	×11

D Multiply by eleven:

three hundred and sixty-seven

five hundred and seven

one hundred and forty

Multiplication by 12

A

$4 \times 12 = \square$	$9 \times 12 = \square$	$5 \times 12 = \square$	$3 \times 12 = \square$	$1 \times 12 = \square$
$0 \times 12 = \square$	$2 \times 12 = \square$	$8 \times 12 = \square$	$7 \times 12 = \square$	$6 \times 12 = \square$
$12 \times \square = 12$	$12 \times \square = 60$	$12 \times \square = 84$	$12 \times \square = 36$	$12 \times \square = 108$
$12 \times \square = 48$	$12 \times \square = 72$	$12 \times \square = 24$	$12 \times \square = 0$	$12 \times \square = 96$

B Add:

$36 + 5 = \square$	$12 + 9 = \square$	$108 + 3 = \square$	$72 + 8 = \square$	$36 + 6 = \square$
$108 + 4 = \square$	$84 + 8 = \square$	$24 + 7 = \square$	$108 + 2 = \square$	$48 + 5 = \square$
$96 + 4 = \square$	$48 + 3 = \square$	$24 + 6 = \square$	$48 + 2 = \square$	$96 + 6 = \square$
$96 + 5 = \square$	$48 + 4 = \square$	$24 + 8 = \square$	$96 + 8 = \square$	$48 + 7 = \square$
$108 + 8 = \square$	$12 + 8 = \square$	$108 + 7 = \square$	$72 + 9 = \square$	$36 + 4 = \square$
$84 + 9 = \square$	$48 + 8 = \square$	$36 + 8 = \square$	$24 + 9 = \square$	$108 + 6 = \square$
$108 + 9 = \square$	$96 + 9 = \square$	$84 + 6 = \square$	$48 + 6 = \square$	$96 + 7 = \square$
$36 + 9 = \square$	$48 + 9 = \square$	$84 + 7 = \square$	$108 + 5 = \square$	$36 + 7 = \square$

C

726	450	375	165	270	802
×12	×12	×12	×12	×12	×12

D

479×12	729×12	285×12	651×12
163×12	455×12	400×12	394×12
821×12	666×12		

E Find the product of:

430 and 12

636 and 12

521 and 12

Mutiply by 12: 386 346 409 814 707

Multiplication

A

47 ×5	56 ×7	89 ×4	32 ×3	91 ×5	70 ×6	68 ×2
95 ×7	43 ×9	55 ×8	80 ×12	39 ×9	67 ×11	75 ×8
172 ×8	698 ×3	534 ×11	901 ×2	508 ×12	476 ×6	
202 ×4	572 ×9	398 ×7	641 ×7	920 ×6	731 ×5	
4127 ×2	3048 ×3	1472 ×5	1731 ×4	1079 ×7		
1168 ×8	1034 ×9	1006 ×7	3541 ×2	2345 ×3		

B Solve by multiplication:

58 + 58 + 58 + 58 + 58 + 58 + 58 + 58

427 + 427 + 427 + 427

1027 + 1027 + 1027 + 1027

698 + 698 + 698 + 698 + 698

C Find the product of:

355 and 9   1076 and 8

463 and 12   2047 and 4

D Multiply three thousand and six by three.

What number is six times ninety-seven?

What number is four times three hundred and ten?

Find the product of   1247 and 4   1309 and 6   1099 and 8

Multiply one thousand and twelve by seven.

Division

A

$2\overline{)666}$	$3\overline{)999}$	$4\overline{)888}$	$6\overline{)666}$	$5\overline{)555}$	$2\overline{)444}$
$2\overline{)184}$	$3\overline{)153}$	$4\overline{)244}$	$5\overline{)255}$	$6\overline{)186}$	$5\overline{)355}$
$3\overline{)516}$	$4\overline{)768}$	$5\overline{)755}$	$6\overline{)966}$	$4\overline{)644}$	$3\overline{)543}$
$3\overline{)575}$	$4\overline{)726}$	$5\overline{)957}$	$6\overline{)849}$	$6\overline{)788}$	$5\overline{)657}$
$4\overline{)852}$	$5\overline{)595}$	$4\overline{)476}$	$6\overline{)684}$	$3\overline{)981}$	$2\overline{)478}$
$3\overline{)861}$	$6\overline{)972}$	$5\overline{)780}$	$2\overline{)972}$	$6\overline{)864}$	$4\overline{)776}$
$5\overline{)540}$	$4\overline{)836}$	$5\overline{)545}$	$3\overline{)927}$	$6\overline{)654}$	$2\overline{)818}$
$6\overline{)675}$	$2\overline{)579}$	$5\overline{)748}$	$3\overline{)763}$	$4\overline{)975}$	$3\overline{)826}$
$2\overline{)340}$	$5\overline{)750}$	$6\overline{)840}$	$4\overline{)760}$	$6\overline{)960}$	$5\overline{)650}$
$2\overline{)660}$	$4\overline{)840}$	$3\overline{)690}$	$5\overline{)550}$	$3\overline{)390}$	$2\overline{)820}$
$3\overline{)600}$	$4\overline{)800}$	$2\overline{)400}$	$3\overline{)900}$	$3\overline{)600}$	$2\overline{)800}$
$4\overline{)707}$	$5\overline{)708}$	$3\overline{)704}$	$6\overline{)904}$	$6\overline{)805}$	$5\overline{)807}$
$4\overline{)790}$	$5\overline{)473}$	$3\overline{)502}$	$2\overline{)670}$	$6\overline{)529}$	$4\overline{)902}$

B

$476 \div 5$ $396 \div 6$ $854 \div 4$

$839 \div 2$ $765 \div 3$ $929 \div 4$

C

Divide 470 by 4.

How many 5s in 543?

Share 867 by 6.

Share 973 by 3.

How many 2s in 763?

Divide 374 by 4.

Division

A

$2\overline{)6842}$ $4\overline{)8448}$ $3\overline{)6939}$ $2\overline{)8246}$ $3\overline{)3963}$

$3\overline{)1254}$ $6\overline{)1866}$ $2\overline{)1284}$ $5\overline{)1555}$ $4\overline{)1684}$

$4\overline{)3649}$ $5\overline{)4556}$ $3\overline{)2168}$ $2\overline{)1664}$ $6\overline{)3667}$

$6\overline{)7868}$ $4\overline{)6848}$ $5\overline{)8556}$ $2\overline{)7886}$ $3\overline{)8167}$

$5\overline{)8657}$ $2\overline{)7563}$ $3\overline{)8258}$ $6\overline{)7569}$ $4\overline{)9567}$

$2\overline{)3273}$ $6\overline{)1572}$ $3\overline{)5264}$ $5\overline{)7845}$ $4\overline{)6693}$

$5\overline{)5473}$ $2\overline{)2176}$ $4\overline{)4372}$ $3\overline{)9251}$ $6\overline{)6583}$

$6\overline{)9645}$ $4\overline{)7632}$ $3\overline{)7521}$ $2\overline{)1812}$ $5\overline{)6538}$

$4\overline{)6563}$ $2\overline{)9761}$ $6\overline{)8823}$ $3\overline{)5852}$ $5\overline{)7854}$

$6\overline{)5822}$ $2\overline{)1818}$ $3\overline{)6244}$ $4\overline{)8273}$ $5\overline{)4543}$

$2\overline{)8035}$ $4\overline{)7205}$ $3\overline{)9048}$ $6\overline{)8406}$ $5\overline{)3505}$

$2\overline{)4015}$ $5\overline{)5025}$ $3\overline{)6027}$ $6\overline{)6054}$ $4\overline{)8034}$

B Divide 1247 by 2.

Divide 5037 by 5.

Share 7603 by 6.

Share 8321 by 4.

C How many threes in three thousand and six?

How many times can six be taken from four thousand and eighty-six?

Share six thousand into four equal parts.

Division by 7

A

$14 \div 7 = \square$ $49 \div 7 = \square$ $21 \div 7 = \square$ $63 \div 7 = \square$ $0 \div 7 = \square$

$35 \div 7 = \square$ $56 \div 7 = \square$ $7 \div 7 = \square$ $28 \div 7 = \square$ $42 \div 7 = \square$

B

$62 - 56 = \square$ $41 - 35 = \square$ $20 - 14 = \square$ $54 - 49 = \square$ $33 - 28 = \square$

$13 - 7 = \square$ $60 - 56 = \square$ $52 - 49 = \square$ $30 - 28 = \square$ $55 - 49 = \square$

$32 - 28 = \square$ $61 - 56 = \square$ $50 - 49 = \square$ $12 - 7 = \square$ $40 - 35 = \square$

$53 - 49 = \square$ $11 - 7 = \square$ $51 - 49 = \square$ $34 - 28 = \square$ $31 - 28 = \square$

C

$7\overline{)55}$ $7\overline{)53}$ $7\overline{)12}$ $7\overline{)61}$ $7\overline{)32}$

$7\overline{)52}$ $7\overline{)30}$ $7\overline{)62}$ $7\overline{)20}$ $7\overline{)33}$

$7\overline{)11}$ $7\overline{)60}$ $7\overline{)51}$ $7\overline{)13}$ $7\overline{)40}$

$7\overline{)54}$ $7\overline{)34}$ $7\overline{)50}$ $7\overline{)41}$ $7\overline{)31}$

D

$7\overline{)95}$ $7\overline{)87}$ $7\overline{)79}$ $7\overline{)83}$ $7\overline{)92}$ $7\overline{)78}$ $7\overline{)89}$

$7\overline{)135}$ $7\overline{)146}$ $7\overline{)164}$ $7\overline{)151}$ $7\overline{)182}$ $7\overline{)173}$

$7\overline{)943}$ $7\overline{)408}$ $7\overline{)567}$ $7\overline{)680}$ $7\overline{)884}$ $7\overline{)362}$

$7\overline{)1543}$ $7\overline{)1739}$ $7\overline{)1962}$ $7\overline{)1808}$ $7\overline{)1650}$

$7\overline{)6306}$ $7\overline{)3435}$ $7\overline{)4389}$ $7\overline{)5928}$ $7\overline{)2769}$

$7\overline{)8607}$ $7\overline{)7350}$ $7\overline{)8846}$ $7\overline{)9429}$ $7\overline{)9707}$

E Divide by seven:

4271 909 4210 1435 892

Division by 8

A

$8\overline{)24}$	$8\overline{)72}$	$8\overline{)0}$	$8\overline{)64}$	$8\overline{)16}$
$8\overline{)48}$	$8\overline{)8}$	$8\overline{)56}$	$8\overline{)32}$	$8\overline{)40}$

B

$63 - 56 = \square$	$23 - 16 = \square$	$54 - 48 = \square$	$71 - 64 = \square$	$31 - 24 = \square$
$15 - 8 = \square$	$61 - 56 = \square$	$21 - 16 = \square$	$11 - 8 = \square$	$55 - 48 = \square$
$70 - 64 = \square$	$51 - 48 = \square$	$13 - 8 = \square$	$62 - 56 = \square$	$53 - 48 = \square$
$14 - 8 = \square$	$60 - 56 = \square$	$30 - 24 = \square$	$12 - 8 = \square$	$20 - 16 = \square$
$10 - 8 = \square$	$52 - 48 = \square$	$22 - 16 = \square$	$50 - 48 = \square$	

C

$15 \div 8$	$63 \div 8$	$52 \div 8$	$20 \div 8$	$71 \div 8$	$30 \div 8$
$12 \div 8$	$54 \div 8$	$60 \div 8$	$11 \div 8$	$53 \div 8$	$70 \div 8$
$23 \div 8$	$10 \div 8$	$55 \div 8$	$62 \div 8$	$22 \div 8$	$50 \div 8$
$51 \div 8$	$14 \div 8$	$61 \div 8$	$21 \div 8$	$13 \div 8$	$31 \div 8$

D

$8\overline{)95}$	$8\overline{)87}$	$8\overline{)96}$	$8\overline{)88}$	$8\overline{)93}$	$8\overline{)98}$	$8\overline{)89}$
$8\overline{)147}$	$8\overline{)186}$	$8\overline{)192}$	$8\overline{)142}$	$8\overline{)106}$	$8\overline{)155}$	
$8\overline{)946}$	$8\overline{)972}$	$8\overline{)806}$	$8\overline{)909}$	$8\overline{)992}$	$8\overline{)958}$	
$8\overline{)1962}$	$8\overline{)1179}$	$8\overline{)1894}$	$8\overline{)1758}$	$8\overline{)1632}$		
$8\overline{)7506}$	$8\overline{)6330}$	$8\overline{)5749}$	$8\overline{)7072}$	$8\overline{)4752}$		
$8\overline{)8903}$	$8\overline{)9243}$	$8\overline{)9071}$	$8\overline{)8749}$	$8\overline{)9673}$		

E

$4320 \div 8$	$9041 \div 8$	$729 \div 8$	$1690 \div 8$

Division by 9

A

$\square \div 9 = 6$ $\quad$ $\square \div 9 = 9$ $\quad$ $\square \div 9 = 2$ $\quad$ $\square \div 9 = 4$ $\quad$ $\square \div 9 = 7$

$\square \div 9 = 0$ $\quad$ $\square \div 9 = 5$ $\quad$ $\square \div 9 = 1$ $\quad$ $\square \div 9 = 3$ $\quad$ $\square \div 9 = 8$

B

$9\overline{)70}$ $\quad$ $9\overline{)26}$ $\quad$ $9\overline{)53}$ $\quad$ $9\overline{)33}$ $\quad$ $9\overline{)23}$ $\quad$ $9\overline{)42}$ $\quad$ $9\overline{)20}$

$9\overline{)30}$ $\quad$ $9\overline{)12}$ $\quad$ $9\overline{)62}$ $\quad$ $9\overline{)43}$ $\quad$ $9\overline{)80}$ $\quad$ $9\overline{)50}$ $\quad$ $9\overline{)60}$

$9\overline{)13}$ $\quad$ $9\overline{)34}$ $\quad$ $9\overline{)21}$ $\quad$ $9\overline{)11}$ $\quad$ $9\overline{)32}$ $\quad$ $9\overline{)22}$ $\quad$ $9\overline{)16}$

$9\overline{)52}$ $\quad$ $9\overline{)61}$ $\quad$ $9\overline{)51}$ $\quad$ $9\overline{)24}$ $\quad$ $9\overline{)71}$ $\quad$ $9\overline{)41}$ $\quad$ $9\overline{)40}$

$9\overline{)14}$ $\quad$ $9\overline{)25}$ $\quad$ $9\overline{)44}$ $\quad$ $9\overline{)35}$ $\quad$ $9\overline{)15}$ $\quad$ $9\overline{)17}$ $\quad$ $9\overline{)31}$

C

$9\overline{)106}$ $\quad$ $9\overline{)123}$ $\quad$ $9\overline{)144}$ $\quad$ $9\overline{)192}$ $\quad$ $9\overline{)137}$ $\quad$ $9\overline{)187}$

$9\overline{)175}$ $\quad$ $9\overline{)146}$ $\quad$ $9\overline{)181}$ $\quad$ $9\overline{)110}$ $\quad$ $9\overline{)199}$ $\quad$ $9\overline{)168}$

$9\overline{)619}$ $\quad$ $9\overline{)820}$ $\quad$ $9\overline{)332}$ $\quad$ $9\overline{)264}$ $\quad$ $9\overline{)676}$ $\quad$ $9\overline{)488}$

$9\overline{)563}$ $\quad$ $9\overline{)401}$ $\quad$ $9\overline{)529}$ $\quad$ $9\overline{)255}$ $\quad$ $9\overline{)767}$ $\quad$ $9\overline{)309}$

$9\overline{)1301}$ $\quad$ $9\overline{)1274}$ $\quad$ $9\overline{)1165}$ $\quad$ $9\overline{)1750}$ $\quad$ $9\overline{)1639}$

$9\overline{)8193}$ $\quad$ $9\overline{)1083}$ $\quad$ $9\overline{)1416}$ $\quad$ $9\overline{)7657}$ $\quad$ $9\overline{)1848}$

$9\overline{)6039}$ $\quad$ $9\overline{)2300}$ $\quad$ $9\overline{)3945}$ $\quad$ $9\overline{)5466}$ $\quad$ $9\overline{)6578}$

D Share by nine:

376 $\quad$ 4207 $\quad$ 3943 $\quad$ 680

5279 $\quad$ 809 $\quad$ 976 $\quad$ 8280

How many groups of nine in:

476? $\quad$ 2146? $\quad$ 1906? $\quad$ 694? $\quad$ 923?

Division by 11

A

$33 \div 11 = \square$ $66 \div 11 = \square$ $88 \div 11 = \square$ $11 \div 11 = \square$ $99 \div 11 = \square$

$77 \div 11 = \square$ $55 \div 11 = \square$ $0 \div 11 = \square$ $44 \div 11 = \square$ $22 \div 11 = \square$

B

$20 - 11 = \square$ $97 - 88 = \square$ $84 - 77 = \square$ $52 - 44 = \square$ $42 - 33 = \square$

$81 - 77 = \square$ $73 - 66 = \square$ $41 - 33 = \square$ $94 - 88 = \square$ $64 - 55 = \square$

$96 - 88 = \square$ $62 - 55 = \square$ $30 - 22 = \square$ $75 - 66 = \square$ $40 - 33 = \square$

$71 - 66 = \square$ $53 - 44 = \square$ $92 - 88 = \square$ $63 - 55 = \square$ $51 - 44 = \square$

$93 - 88 = \square$ $86 - 77 = \square$ $72 - 66 = \square$ $50 - 44 = \square$ $95 - 88 = \square$

$80 - 77 = \square$ $91 - 88 = \square$ $70 - 66 = \square$ $31 - 22 = \square$ $83 - 77 = \square$

$61 - 55 = \square$ $82 - 77 = \square$ $60 - 55 = \square$ $85 - 77 = \square$ $74 - 66 = \square$

C

$11\overline{)150}$ $11\overline{)141}$ $11\overline{)173}$ $11\overline{)192}$ $11\overline{)165}$ $11\overline{)186}$

$11\overline{)674}$ $11\overline{)590}$ $11\overline{)869}$ $11\overline{)482}$ $11\overline{)334}$ $11\overline{)752}$

$11\overline{)103}$ $11\overline{)105}$ $11\overline{)109}$ $11\overline{)108}$ $11\overline{)107}$ $11\overline{)106}$

$11\overline{)1607}$ $11\overline{)1983}$ $11\overline{)1736}$ $11\overline{)1870}$ $11\overline{)1525}$

$11\overline{)7249}$ $11\overline{)2207}$ $11\overline{)8125}$ $11\overline{)9936}$ $11\overline{)4510}$

$11\overline{)1074}$ $11\overline{)1029}$ $11\overline{)1030}$ $11\overline{)1056}$ $11\overline{)1063}$

D How many groups of eleven in:

3432? 605? 5841? 352?

6270? 9174? 308? 902?

Divide by 11:

6294 942 2201 814 1098

Division by 12

A

$12\overline{)108}$	$12\overline{)24}$	$12\overline{)36}$	$12\overline{)72}$	$12\overline{)12}$
$12\overline{)60}$	$12\overline{)84}$	$12\overline{)96}$	$12\overline{)0}$	$12\overline{)48}$

B

$12\overline{)43}$	$12\overline{)56}$	$12\overline{)33}$	$12\overline{)104}$	$12\overline{)81}$	$12\overline{)101}$
$12\overline{)90}$	$12\overline{)42}$	$12\overline{)82}$	$12\overline{)102}$	$12\overline{)52}$	$12\overline{)105}$
$12\overline{)20}$	$12\overline{)91}$	$12\overline{)51}$	$12\overline{)100}$	$12\overline{)81}$	$12\overline{)103}$
$12\overline{)55}$	$12\overline{)41}$	$12\overline{)21}$	$12\overline{)50}$	$12\overline{)57}$	$12\overline{)58}$
$12\overline{)93}$	$12\overline{)94}$	$12\overline{)53}$	$12\overline{)46}$	$12\overline{)92}$	$12\overline{)30}$
$12\overline{)32}$	$12\overline{)45}$	$12\overline{)54}$	$12\overline{)40}$	$12\overline{)80}$	$12\overline{)44}$

C

$12\overline{)749}$	$12\overline{)684}$	$12\overline{)379}$	$12\overline{)364}$	$12\overline{)851}$	$12\overline{)929}$
$12\overline{)478}$	$12\overline{)607}$	$12\overline{)963}$	$12\overline{)729}$	$12\overline{)216}$	$12\overline{)414}$
$12\overline{)113}$	$12\overline{)111}$	$12\overline{)112}$	$12\overline{)117}$	$12\overline{)110}$	$12\overline{)115}$

$12\overline{)2046}$	$12\overline{)2301}$	$12\overline{)2192}$	$12\overline{)2210}$	$12\overline{)1937}$
$12\overline{)9704}$	$12\overline{)7530}$	$12\overline{)6760}$	$12\overline{)2921}$	$12\overline{)4785}$
$12\overline{)1165}$	$12\overline{)1024}$	$12\overline{)1072}$	$12\overline{)1193}$	$12\overline{)1150}$

D

$7216 \div 12$	$3406 \div 12$	$808 \div 12$
$7206 \div 12$	$4200 \div 12$	$8833 \div 12$
$1111 \div 12$	$404 \div 12$	

E How many groups of twelve in:

2076? 986? 4270? 6735? 406?

Division

A

$2\overline{)93}$ $12\overline{)97}$ $4\overline{)63}$ $3\overline{)85}$ $7\overline{)90}$ $10\overline{)90}$

$12\overline{)89}$ $8\overline{)75}$ $6\overline{)49}$ $11\overline{)98}$ $5\overline{)39}$ $9\overline{)71}$

$2\overline{)730}$ $3\overline{)745}$ $7\overline{)962}$ $6\overline{)854}$ $4\overline{)918}$ $5\overline{)870}$

$5\overline{)195}$ $7\overline{)189}$ $3\overline{)163}$ $4\overline{)158}$ $2\overline{)137}$ $6\overline{)176}$

$12\overline{)837}$ $7\overline{)654}$ $11\overline{)439}$ $8\overline{)546}$ $9\overline{)752}$ $10\overline{)639}$

$5\overline{)7472}$ $3\overline{)5914}$ $2\overline{)5345}$ $4\overline{)7654}$ $6\overline{)8254}$

$11\overline{)1669}$ $12\overline{)1954}$ $9\overline{)1743}$ $10\overline{)1265}$ $8\overline{)1943}$

$7\overline{)6743}$ $5\overline{)4536}$ $8\overline{)7325}$ $6\overline{)9435}$ $4\overline{)3726}$

$4\overline{)1228}$ $5\overline{)2350}$ $7\overline{)6342}$ $8\overline{)6640}$ $12\overline{)8400}$

$4\overline{)4028}$ $5\overline{)6520}$ $8\overline{)9840}$ $9\overline{)9468}$ $3\overline{)9018}$

B How many twelves make three hundred and forty-eight?

Share two thousand and eight by three.

Divide four thousand and sixty by seven.

How many groups of eight in two thousand?

What must be added to one thousand and fifteen so that it will divide equally by seven?

C $4206 \div 8$ $379 \div 4$ $1463 \div 11$ $909 \div 5$

$8080 \div 12$ $600 \div 7$ $3607 \div 6$ $627 \div 9$

Notation

A Write in words the value of the 7 in these numbers.

7463 2765 2617 5670 2007

4799 8673 7421 3740 7621

B Rearrange each group of figures to make the largest number possible.

4516 8103 4973 9624 1003

1077 9427 1083 3562 7462

C Multiply each number by 10 and write its value in words.

362 75 9 81 427 808

Multiply each number by 100 and write its value in words.

5 46 50 8 96 72

7 12

D Divide each number by 10 and write its value in words.

370 40 4270 6320 580 70

Divide each number by 100 and write its value in words.

3400 300 7100 800 5200

E Write what you must do to make:

7 become 70. 340 become 34.

4000 become 40. 50 become 500.

20 become 2000. 60 become 6.

F Find the value of X in the following.

$550 = X \times 10$ $4500 = X \times 10$ $1300 = X \times 100$

$1020 = X \times 10$ $630 = X \times 10$ $1450 = X \times 10$

$41 = X \div 100$ $6 = X \div 100$ $92 = X \div 10$

$320 = X \div 10$ $104 = X \div 10$ $16 = X \div 100$

Number series and equations

A Write out and complete the number series.

4, 8, □, 16, 20, □, 28, □

9, 12, □, □, 21, 24, □

81, 72, 63, □, □, 36, □

50, □, 70, □, 90, □

24, 22, 20, □, □, □

□, 10, 15, □, □, □, 35

66, □, 44, 33, □, □

24, 32, □, □, 56, 64, □

48, 42, □, 30, □, 18, □

7, 14, 21, 28, □, □, □

48, □, 72, 84, 96, □, □

88, 80, 72, □, □, □

□, 14, 16, □, 20, 22, □

□, 90, □, 70, □, 50, □

32, □, 24, □, 16, 12, □

55, 66, 77, □, □, □

□, 27, 36, □, □, 63, 72

77, □, 63, □, 49, □, 35

B Write out and complete these number series.

3, 6, 10, 13, 17, □, □, □, □

52, 47, 39, 34, 26, □, □, □, □

2, 8, 10, 16, 18, □, □, □, □

62, 61, 55, 54, 48, □, □, □, □

4, 11, 19, 26, 34, □, □, □, □

24, 22, 19, 17, 14, □, □, □, □

8, 13, 22, 27, 36, □, □, □, □

54, 50, 43, 39, 32, □, □, □, □

C Write and complete the equations.

$6 \times 2 = \square \times 4$

$\square \times 8 = 4 \times 12$

$10 \times 4 = 8 \times \square$

$8 \times \square = 12 \times 2$

$\square \times 9 = 6 \times 6$

$12 \times 5 = \square \times 10$

$2 \times 10 = 4 \times \square$

$\square \times 8 = 4 \times 4$

$12 \times \square = 2 \times 6$

$4 \times 3 = \square \times 6$

$9 \times 4 = \square \times 6$

$6 \times \square = 3 \times 8$

$8 \times \square = 4 \times 12$

$12 \times 4 = \square \times 8$

$5 \times 12 = \square \times 6$

$3 \times 8 = \square \times 12$

$\square \times 10 = 8 \times 5$

$8 \times 3 = \square \times 4$

$4 \times 6 = \square \times 3$

$2 \times \square = 4 \times 6$

$10 \times \square = 8 \times 5$

$3 \times 4 = \square \times 6$

$\square \times 12 = 6 \times 2$

$10 \times 4 = \square \times 8$

$8 \times 6 = \square \times 4$

$12 \times 2 = \square \times 6$

$5 \times \square = 2 \times 10$

Fractions

A What fraction of each shape is: **a** shaded? **b** unshaded?

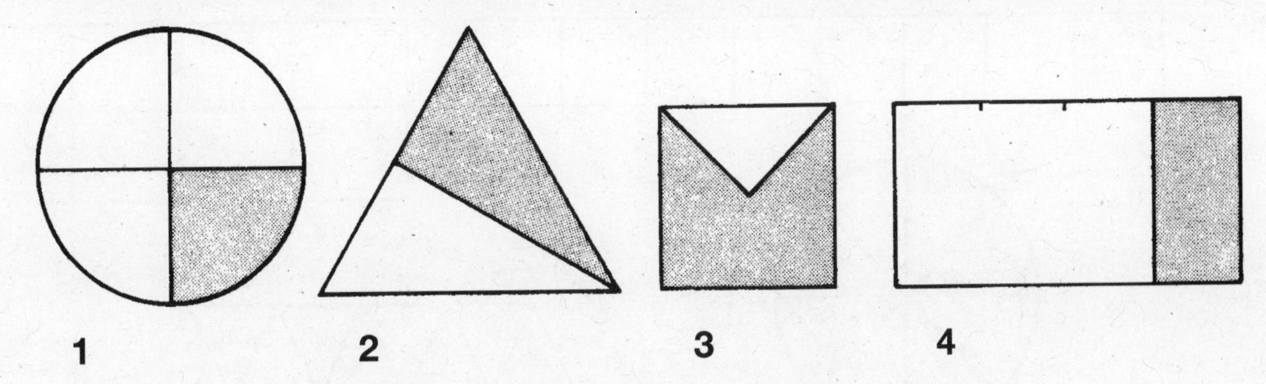

B

A

B

C

D

Which line is $\frac{1}{2}$ of B?

Which line is $\frac{1}{4}$ of A?

Which line is $\frac{1}{2}$ of A?

Which line is $\frac{3}{4}$ of A?

C How many cm is:

$\frac{1}{4}$ of A? $\frac{1}{2}$ of C? $\frac{1}{2}$ of B?

$\frac{1}{2}$ of D? $\frac{1}{4}$ of B? $\frac{3}{4}$ of A?

D Solve:

$\frac{1}{2}$ of 6p $\frac{1}{4}$ of 8 $\frac{3}{4}$ of 12p $\frac{1}{2}$ of 20

$\frac{1}{4}$ of 16 $\frac{1}{2}$ of 24p $\frac{1}{4}$ of 20 $\frac{3}{4}$ of 24

$\frac{1}{2}$ of 40p $\frac{1}{4}$ of 40 $\frac{3}{4}$ of 40 $\frac{1}{2}$ of 18

E Write and complete:

$\frac{3}{4}+\frac{1}{4}=\square$ $\frac{1}{4}+\frac{1}{4}=\square$ $\frac{1}{2}+\frac{1}{4}=\square$ $\frac{1}{4}+\frac{1}{2}+\frac{1}{4}=\square$

$1-\frac{1}{4}=\square$ $\frac{3}{4}-\frac{1}{4}=\square$ $1-\frac{1}{2}=\square$ $\frac{1}{2}-\frac{1}{4}=\square$

$\frac{3}{4}+\frac{3}{4}=\square$ $\frac{1}{2}+\frac{3}{4}=\square$ $1\frac{1}{4}-\frac{1}{2}=\square$ $1\frac{1}{2}-\frac{3}{4}=\square$

Fractions

A What fraction of each shape is: **a** shaded? **b** unshaded?

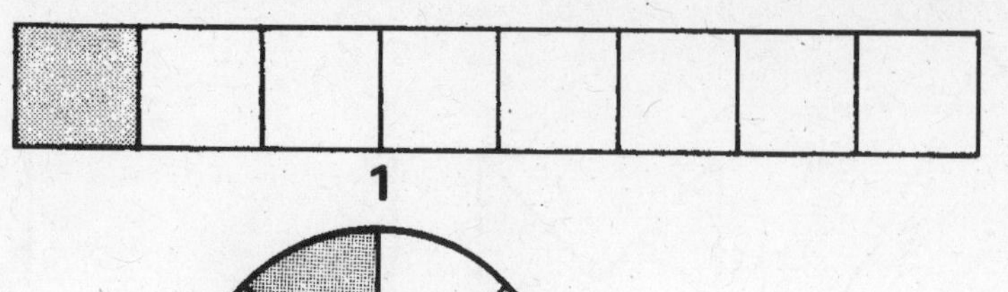

1

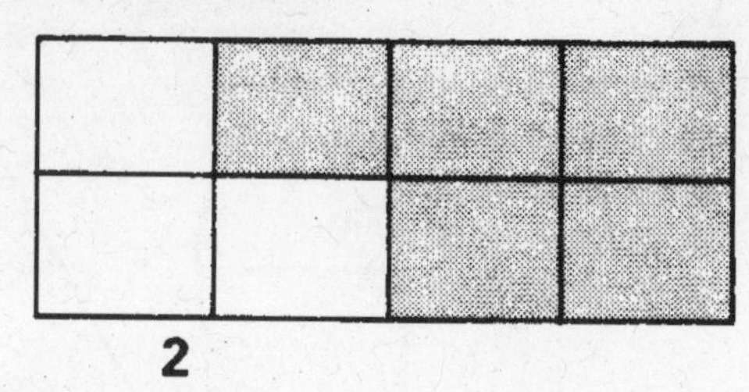

2

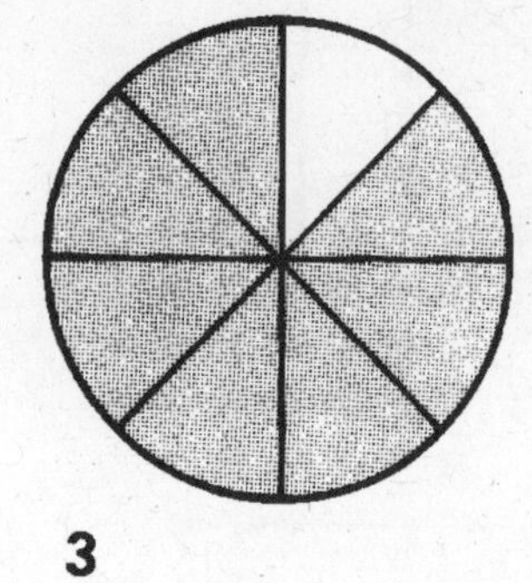

3

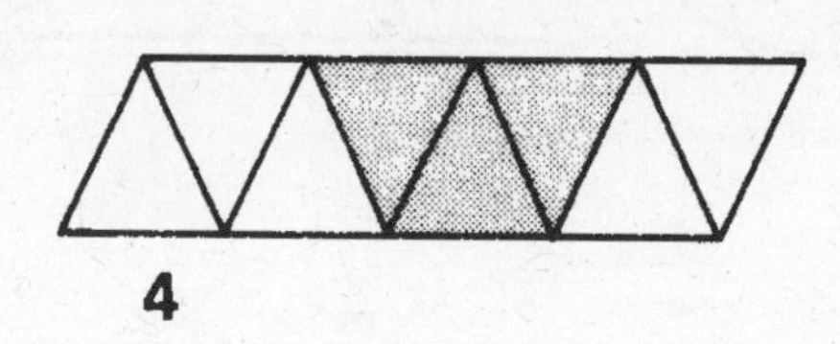

4

B $1 = \frac{\square}{8}$ $3 = \frac{\square}{8}$ $2 = \frac{\square}{8}$ $4 = \frac{\square}{8}$

C $1\frac{5}{8} = \frac{\square}{8}$ $2\frac{3}{8} = \frac{\square}{8}$ $1\frac{1}{8} = \frac{\square}{8}$ $2\frac{5}{8} = \frac{\square}{8}$

D $\frac{1}{4} = \frac{\square}{8}$ $\frac{3}{4} = \frac{\square}{8}$ $\frac{2}{4} = \frac{\square}{8}$ $\frac{1}{2} = \frac{\square}{8}$

E $1\frac{1}{4} = \frac{\square}{8}$ $1\frac{3}{4} = \frac{\square}{8}$ $2\frac{1}{2} = \frac{\square}{8}$ $1\frac{1}{2} = \frac{\square}{8}$

F

The shape is worth 16.
What is the shaded part worth?
What fraction is shaded?

The shape is worth 24.
What is the shaded part worth?
What fraction is shaded?

G Solve:

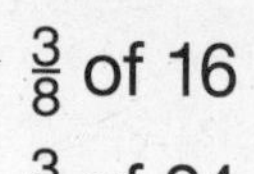

$\frac{3}{8}$ of 16 $\frac{7}{8}$ of 24 $\frac{1}{4}$ of 16 $\frac{3}{4}$ of 16

$\frac{3}{4}$ of 24 $\frac{1}{8}$ of 16 $\frac{5}{8}$ of 24 $\frac{1}{8}$ of 24

$\frac{1}{2}$ of 16 $\frac{1}{4}$ of 24 $\frac{7}{8}$ of 16 $\frac{1}{2}$ of 24

Fractions

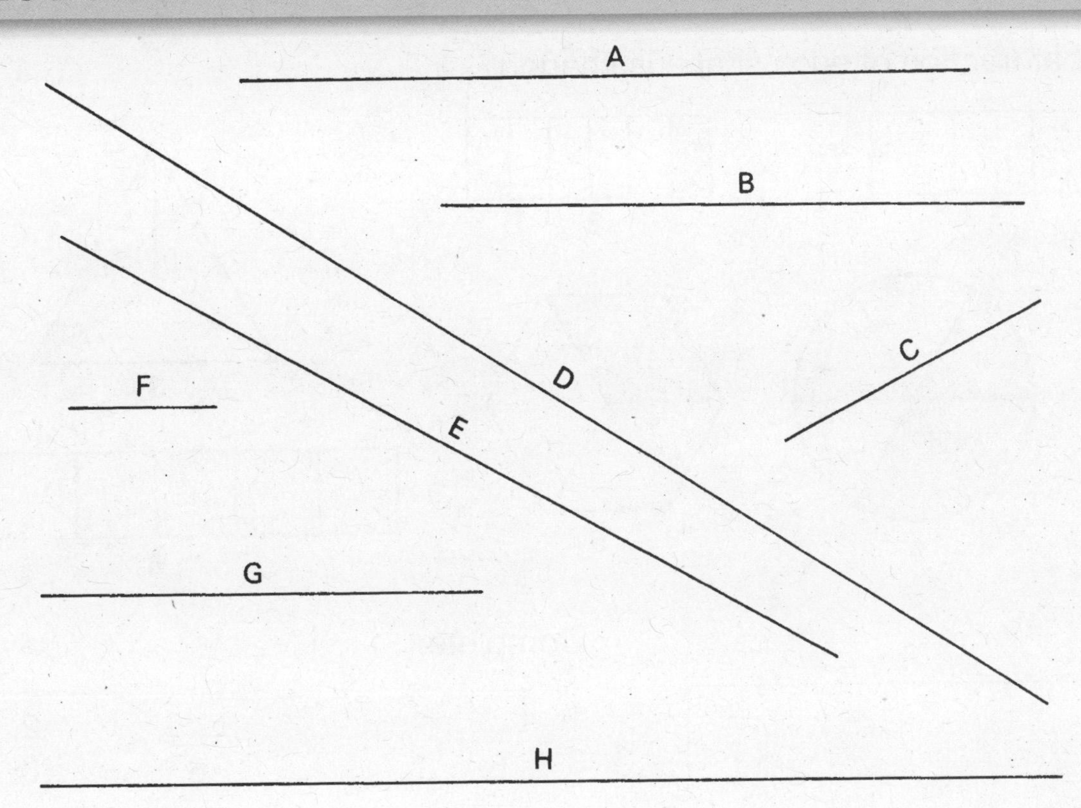

A Which line is:

$\frac{1}{8}$ of D?	$\frac{1}{4}$ of B?	$\frac{5}{8}$ of D?	$\frac{1}{2}$ of C?	$\frac{3}{4}$ of B?
$\frac{1}{2}$ of B?	$\frac{3}{8}$ of D?	$\frac{1}{2}$ of E?	$\frac{7}{8}$ of D?	$\frac{1}{4}$ of D?

B How many cm is:

$\frac{5}{8}$ of D?	$\frac{3}{4}$ of E?	$\frac{1}{2}$ of G?	$\frac{1}{8}$ of B?
$\frac{1}{4}$ of A?	$\frac{3}{8}$ of B?	$\frac{1}{4}$ of C?	$\frac{3}{4}$ of D?
$\frac{1}{2}$ of E?	$\frac{1}{8}$ of D?	$\frac{3}{4}$ of B?	$\frac{1}{4}$ of E?
$\frac{7}{8}$ of B?	$\frac{1}{2}$ of H?	$\frac{1}{2}$ of A?	$\frac{1}{2}$ of F?

C Solve:

$\frac{3}{8}$ of 64p	$\frac{7}{8}$ of 48	$\frac{5}{8}$ of 24 g	$\frac{1}{8}$ of 72
$\frac{5}{8}$ of 24 cm	$\frac{1}{2}$ of 23 cm	$\frac{1}{8}$ of 64 cm	$\frac{1}{4}$ of 22
$\frac{3}{4}$ of 48	$\frac{3}{8}$ of 32p	$\frac{7}{8}$ of 88	$\frac{5}{8}$ of 96 cm
$\frac{1}{8}$ of 56p	$\frac{3}{4}$ of 16	$\frac{3}{8}$ of 40 g	$\frac{7}{8}$ of 80 g

Fractions

A What fraction of each shape is shaded?

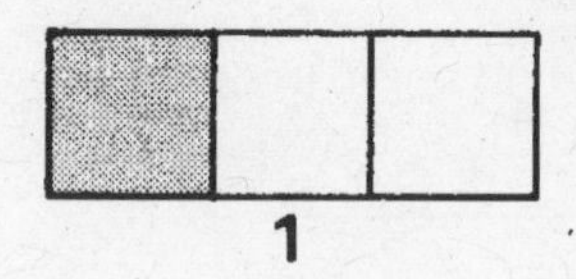

1

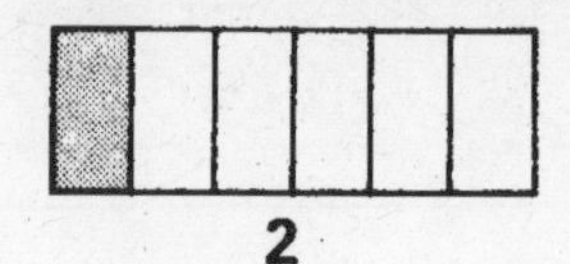

2

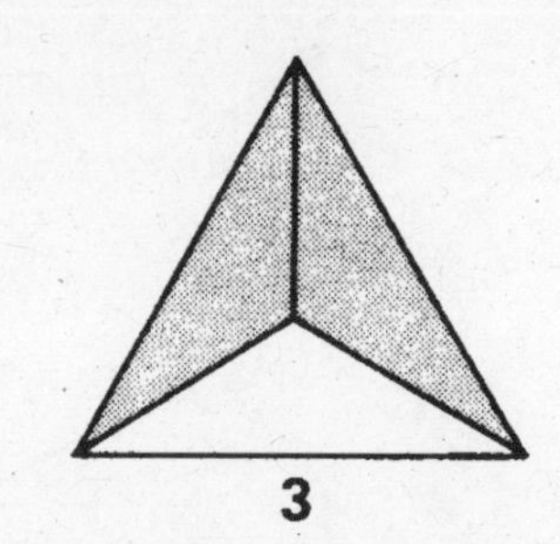

3

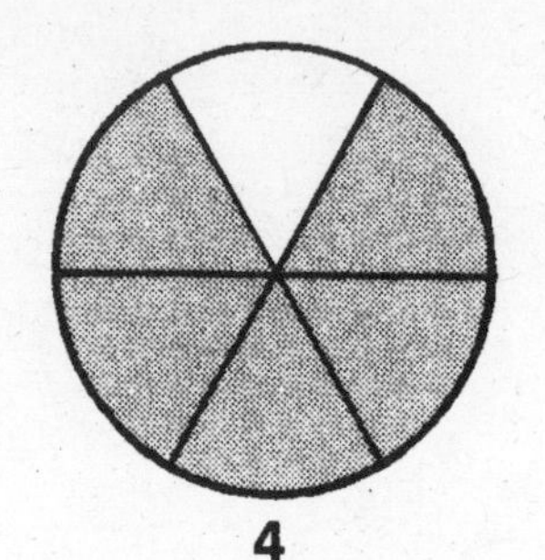

4

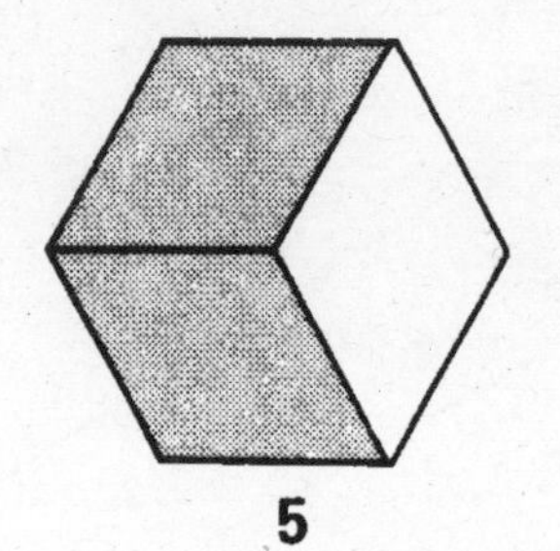

5

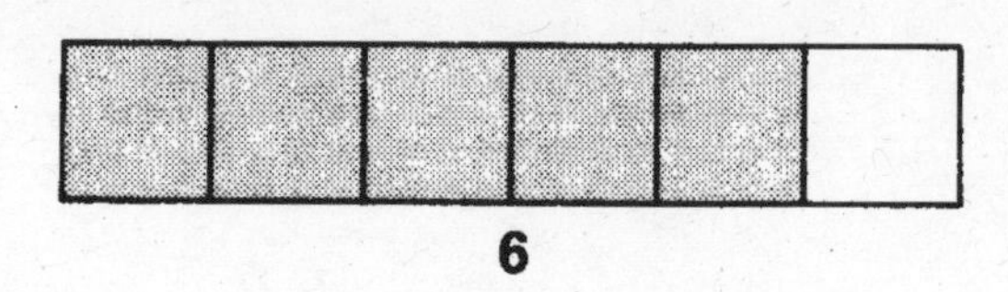

6

B

<table>
<tr><td colspan="6">1</td></tr>
<tr><td colspan="2">$\frac{1}{3}$</td><td colspan="2">$\frac{1}{3}$</td><td colspan="2">$\frac{1}{3}$</td></tr>
<tr><td>$\frac{1}{6}$</td><td>$\frac{1}{6}$</td><td>$\frac{1}{6}$</td><td>$\frac{1}{6}$</td><td>$\frac{1}{6}$</td><td>$\frac{1}{6}$</td></tr>
</table>

Complete:

$\frac{1}{3} = \frac{\square}{6}$ $\quad 1 = \frac{\square}{6}$ $\quad \frac{2}{3} = \frac{\square}{6}$

$1 = \frac{\square}{3}$ $\quad \frac{2}{6} = \frac{\square}{3}$ $\quad \frac{4}{6} = \frac{\square}{3}$

$\frac{\square}{6} = \frac{\square}{3} = 1$

C Write each shaded part in **two** ways.

1 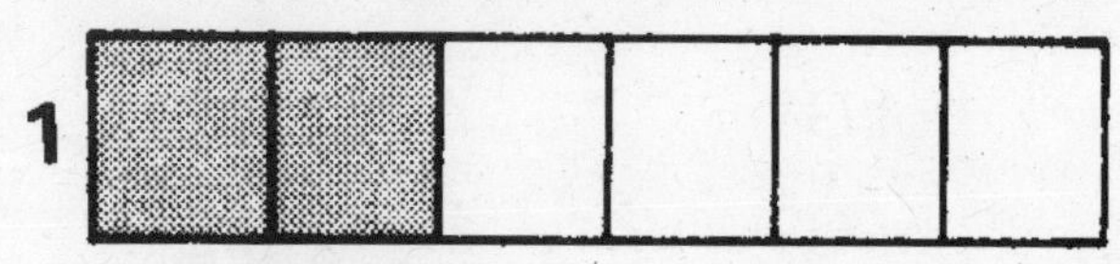 $= \frac{\square}{6} = \frac{\square}{3}$

2 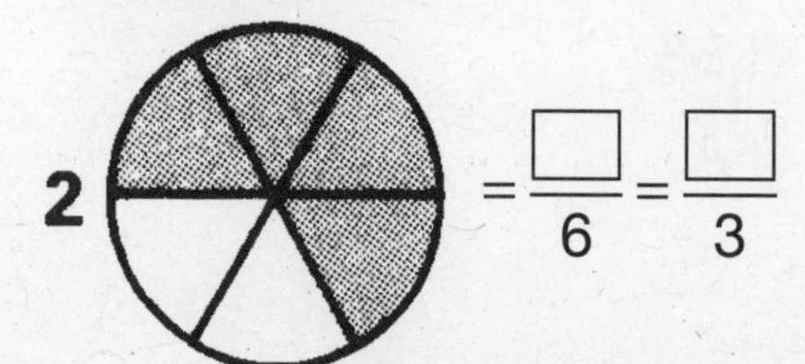 $= \frac{\square}{6} = \frac{\square}{3}$

3 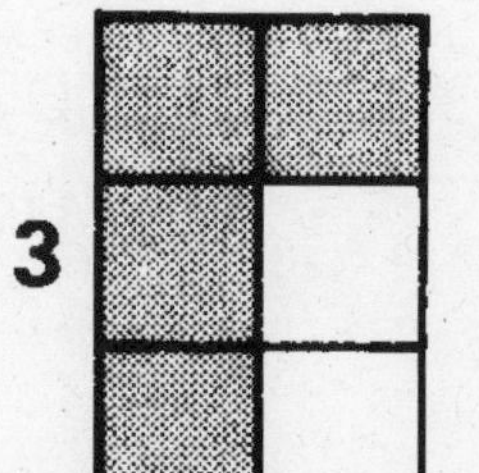 $= \frac{\square}{6} = \frac{\square}{3}$

D

How many are:

$\frac{1}{3}$ of the beads? $\frac{1}{2}$ of the beads? $\frac{1}{6}$ of the beads?

$\frac{5}{6}$ of the beads? $\frac{2}{3}$ of the beads? $\frac{6}{6}$ of the beads?

Fractions

1

$\frac{1}{2}$	$\frac{1}{2}$

$\frac{1}{3}$	$\frac{1}{3}$	$\frac{1}{3}$

$\frac{1}{4}$	$\frac{1}{4}$	$\frac{1}{4}$	$\frac{1}{4}$

$\frac{1}{6}$	$\frac{1}{6}$	$\frac{1}{6}$	$\frac{1}{6}$	$\frac{1}{6}$	$\frac{1}{6}$

$\frac{1}{8}$	$\frac{1}{8}$	$\frac{1}{8}$	$\frac{1}{8}$	$\frac{1}{8}$	$\frac{1}{8}$	$\frac{1}{8}$	$\frac{1}{8}$

A Copy and complete:

$\frac{1}{2} = \frac{\square}{4}$ $\quad \frac{1}{3} = \frac{\square}{6}$ $\quad 1 = \frac{\square}{8}$ $\quad \frac{1}{4} = \frac{\square}{8}$ $\quad 1 = \frac{\square}{6}$

$\frac{3}{4} = \frac{\square}{8}$ $\quad 1 = \frac{\square}{2}$ $\quad \frac{1}{2} = \frac{\square}{6}$ $\quad 1 = \frac{\square}{3}$ $\quad \frac{2}{3} = \frac{\square}{6}$

$1 = \frac{\square}{4}$ $\quad \frac{2}{4} = \frac{\square}{6}$ $\quad \frac{1}{2} = \frac{\square}{8}$ $\quad \frac{2}{4} = \frac{\square}{8}$ $\quad \frac{3}{6} = \frac{\square}{8}$

B

The shape is worth 32.

What fraction is shaded?

What is it worth?

What are the following fractions worth?

$\frac{7}{8}$ $\quad \frac{1}{4}$ $\quad \frac{5}{8}$

$\frac{1}{2}$ $\quad \frac{3}{8}$

C

The shaded part is worth 5.

What is the whole shape worth?

What are the following fractions worth?

$\frac{1}{2}$ $\quad \frac{5}{6}$ $\quad \frac{1}{3}$ $\quad \frac{2}{3}$

Fractions

A What fraction of each shape is: **a** shaded? **b** unshaded?

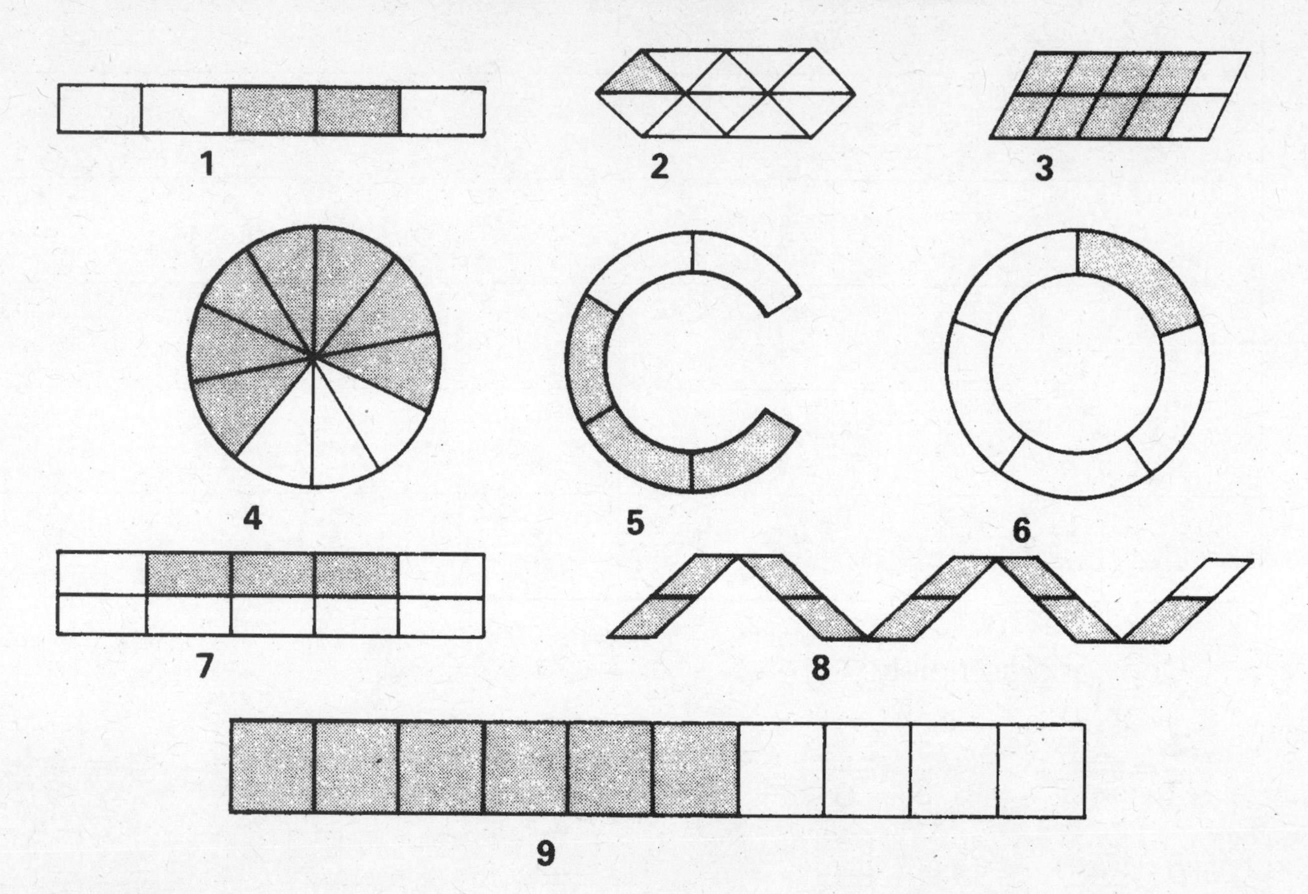

B Complete:

$1 = \frac{\Box}{10}$ $\frac{1}{5} = \frac{\Box}{10}$ $\frac{4}{10} = \frac{\Box}{5}$ $\frac{3}{5} = \frac{\Box}{10}$ $\frac{8}{10} = \frac{\Box}{5}$

$\frac{6}{10} = \frac{\Box}{5}$ $\frac{4}{5} = \frac{\Box}{10}$ $\frac{2}{10} = \frac{\Box}{5}$ $1 = \frac{\Box}{5}$ $\frac{2}{5} = \frac{\Box}{10}$

$\frac{3}{10} + \frac{\Box}{10} = 1$ $\frac{2}{5} + \frac{\Box}{5} = 1$ $\frac{7}{10} + \frac{\Box}{10} = 1$ $\frac{4}{5} + \frac{\Box}{10} = 1$

$\frac{1}{5} + \frac{\Box}{5} = 1$ $\frac{9}{10} + \frac{\Box}{10} = 1$ $\frac{3}{5} + \frac{\Box}{5} = 1$ $\frac{1}{10} + \frac{\Box}{10} = 1$

C Find the value of:

$\frac{1}{5}$ of 25 $\frac{7}{10}$ of 90 cm $\frac{1}{10}$ of 40 g $\frac{3}{5}$ of 75

$\frac{3}{10}$ of 60p $\frac{4}{5}$ of 35 $\frac{9}{10}$ of 20 $\frac{7}{10}$ of 120

$\frac{2}{5}$ of 45 g $\frac{3}{5}$ of 20p $\frac{4}{5}$ of 90 $\frac{9}{10}$ of 30

Fractions

A

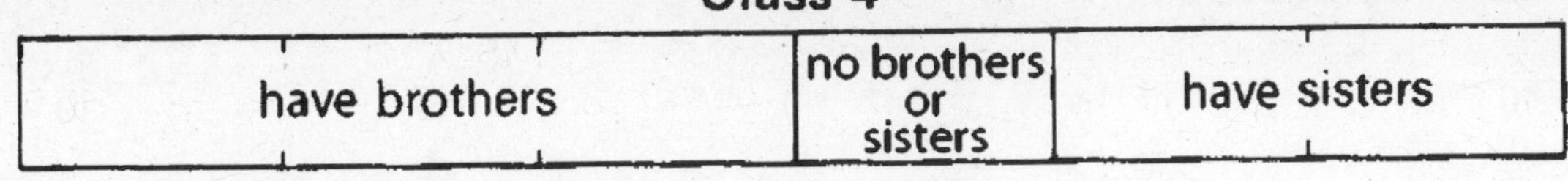

1 What fraction of the class has a brother?

2 What fraction of the class has a sister?

3 What fraction has neither brothers nor sisters?

4 There are 30 children in the class.

How many have: **a** a brother? **b** a sister? **c** neither?

B

1 What fraction has pets?

2 What fraction does not have pets?

3 There are 16 children with no pet. How many have a pet?

C

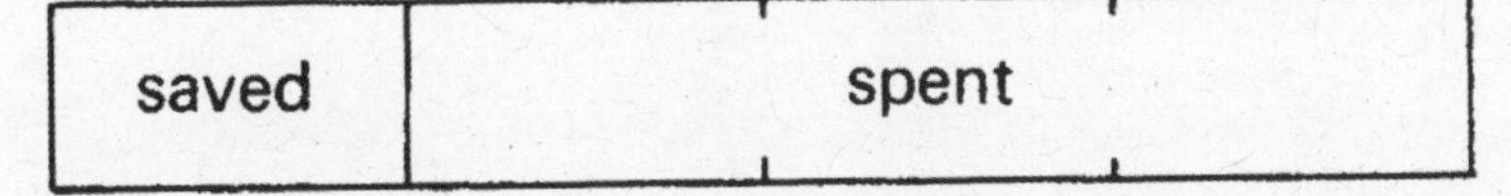

1 Jessica is given 60p each week. What fraction has she saved?

2 What fraction has she spent?

3 How much has she: **a** saved? **b** spent?

D

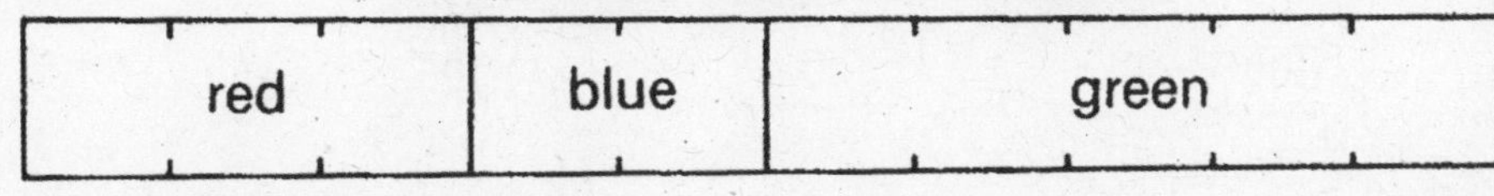

1 What fraction likes: **a** red? **b** blue? **c** green?

2 There are 40 children in the class. How many children like:

a red? **b** blue? **c** green?

Decimal notation

A Write as decimals:

1 $\frac{2}{10}$ $\frac{5}{10}$ $\frac{7}{10}$ $\frac{3}{10}$ $\frac{6}{10}$

$\frac{9}{10}$ $\frac{1}{10}$ $\frac{4}{10}$ $\frac{8}{10}$

2 six tenths, four tenths, three tenths

nine tenths, eight tenths, five tenths

two tenths, seven tenths

3 $4\frac{7}{10}$ $8\frac{3}{10}$ $2\frac{5}{10}$ $7\frac{9}{10}$ $3\frac{4}{10}$

$1\frac{6}{10}$ $8\frac{2}{10}$ $5\frac{8}{10}$ $6\frac{1}{10}$

4 eight point five, nine point one, six point two

three point nine, five point three, seven point four

nine point eight, four point six, two point seven

5 $\frac{15}{100}$ $\frac{23}{100}$ $\frac{67}{100}$ $\frac{21}{100}$

$\frac{19}{100}$ $\frac{83}{100}$ $\frac{52}{100}$ $\frac{32}{100}$

6 $\frac{6}{100}$ $\frac{3}{100}$ $\frac{8}{100}$ $\frac{9}{100}$

$\frac{4}{100}$ $\frac{7}{100}$ $\frac{5}{100}$ $\frac{2}{100}$

$\frac{1}{100}$

7 $\frac{11}{100}$ $\frac{1}{100}$ $\frac{43}{100}$ $\frac{29}{100}$

$\frac{7}{100}$ $\frac{3}{100}$ $\frac{9}{100}$ $\frac{2}{100}$

$\frac{91}{100}$ $\frac{88}{100}$

8 nineteen hundredths, six hundredths

forty-two hundredths, nine hundredths

9 $4\frac{7}{100}$ $9\frac{15}{100}$ $12\frac{37}{100}$ $8\frac{9}{100}$

$11\frac{71}{100}$ $4\frac{3}{100}$

10 four units and six hundredths

nineteen units and fifteen hundredths

five tens, two units and three hundredths

eight units and sixteen hundredths

B Write the value of the underlined figure in words.

$427{\cdot}\underline{6}3$ $43{\cdot}2\underline{6}$ $1\underline{9}{\cdot}2$ $\underline{2}16{\cdot}02$

$219{\cdot}0\underline{6}$ $\underline{3}2{\cdot}47$ $124{\cdot}\underline{2}3$ $1\underline{6}{\cdot}88$

$\underline{1}27{\cdot}14$ $106{\cdot}\underline{2}7$ $324{\cdot}2\underline{9}$ $\underline{4}3{\cdot}44$

Decimal notation

A Write down the greater of each pair of numbers.

0·4 and 0·37	7·02 and 7·1	0·99 and 1·01
4 and 0·47	30·2 and 30·06	18·23 and 18·7
0·02 and 0·11	14·2 and 14·09	

B Arrange each line in order of size, beginning with the greatest.

0·11	1·01	1·10	11·05	11·10	1·11
1·2	1·02	1·22	2·01	1·12	1·01
16·32	16·09	16·23	16·03	16·3	16·2
101·09	101·9	100·99	101·99	100·09	100·9

C Multiply each number by 10.

42·3   42·03   76·27   85·86   143·27

D Multiply each number by 100.

26·03   42·63   18·05   0·2   0·57

E Divide each number by 10.

24·6   91·0   303·1   42·0   271·6

F Divide each number by 100.

3260   421   202   3400   2010

G Write what you must do to make the 7 worth seven tenths in each number.

37·46   73·31   62·37   70·35   7·14

H Write what you must do to each number to make the 5 worth five units.

0·05   53·26   7653   4521·0   62·58

Decimals

A Set down in columns and add.

$4{\cdot}2 + 0{\cdot}02 + 16{\cdot}3$

$124{\cdot}0 + 3{\cdot}44 + 16{\cdot}73$

$102{\cdot}3 + 4{\cdot}27 + 3{\cdot}6$

$12{\cdot}97 + 0{\cdot}09$

$15{\cdot}22 + 101 + 0{\cdot}16$

$4{\cdot}26 + 17 + 92{\cdot}4$

$1{\cdot}2 + 18{\cdot}32 + 216{\cdot}2$

$0{\cdot}07 + 18{\cdot}76 + 4{\cdot}9$

B Set down in columns and subtract.

$18{\cdot}4 - 13{\cdot}26$

$29{\cdot}2 - 15{\cdot}47$

$27{\cdot}08 - 18{\cdot}3$

$99 - 0{\cdot}76$

$54 - 3{\cdot}66$

$42{\cdot}46 - 37{\cdot}09$

$100 - 1{\cdot}07$

$12{\cdot}04 - 0{\cdot}99$

C Subtract the smaller number from the greater.

4·2 and 20

12·63 and 14·2

13·6 and 12·99

4·37 and 18

100 and 94·43

18·63 and 7·6

7·2 and 16·33

0·98 and 105·2

D Set down in columns as decimals and add.

$4\frac{1}{10} + 14\frac{7}{100} + 6\frac{11}{100}$

$26\frac{17}{100} + 32\frac{12}{100} + 12\frac{91}{100}$

$8 + 14\frac{13}{100} + \frac{7}{10}$

$13 + 14\frac{1}{10} + 6\frac{92}{100}$

$12\frac{7}{100} + 15 + 26\frac{13}{100}$

$8\frac{3}{10} + 12\frac{7}{10} + 16\frac{31}{100}$

$\frac{18}{100} + 27\frac{3}{10} + 16\frac{2}{10}$

$12\frac{18}{100} + 47 + \frac{83}{100}$

E Set down in columns as decimals and subtract.

$4\frac{3}{10} - 2\frac{17}{100}$

$100 - \frac{87}{100}$

$18 - 12\frac{18}{100}$

$9\frac{9}{10} - 8\frac{99}{100}$

$14\frac{17}{100} - 7\frac{5}{10}$

$26\frac{3}{10} - 18\frac{19}{100}$

$9\frac{1}{10} - 5\frac{36}{100}$

$42 - 16\frac{7}{100}$

F Fourteen point six minus eight point nine.

Seven point eight four plus nineteen plus six point two.

Find the sum of ninety-six point four and seven point one.

Money – composition to £1·00

A Write down what coins you would receive in change, if you received the least number of coins possible.

cost	coins given	cost	coins given
17p	1 twenty	51p	3 twenties
52p	1 fifty, 1 ten	28p	2 tens, 5 twos
27p	3 tens	69p	7 tens
45p	2 twenties, 3 twos	82p	1 fifty, 2 twenties
21p	5 fives	21p	2 tens, 1 five
62p	1 fifty, 2 tens	33p	3 tens, 2 twos
31p	4 tens	22p	1 ten, 3 fives
41p	3 tens, 3 fives	93p	9 tens, 2 twos

B How much money in each box?

50p	20p	10p	5p	2p	1p
1	1	2	1		

50p	20p	10p	5p	2p	1p
1		2	5	2	

50p	20p	10p	5p	2p	1p
	1	3	6	2	3

50p	20p	10p	5p	2p	1p
	1	3	4	4	2

50p	20p	10p	5p	2p	1p
1	2			3	2

50p	20p	10p	5p	2p	1p
	3	2		7	3

Money – composition to £1·00

A Find the total of each row of coins.

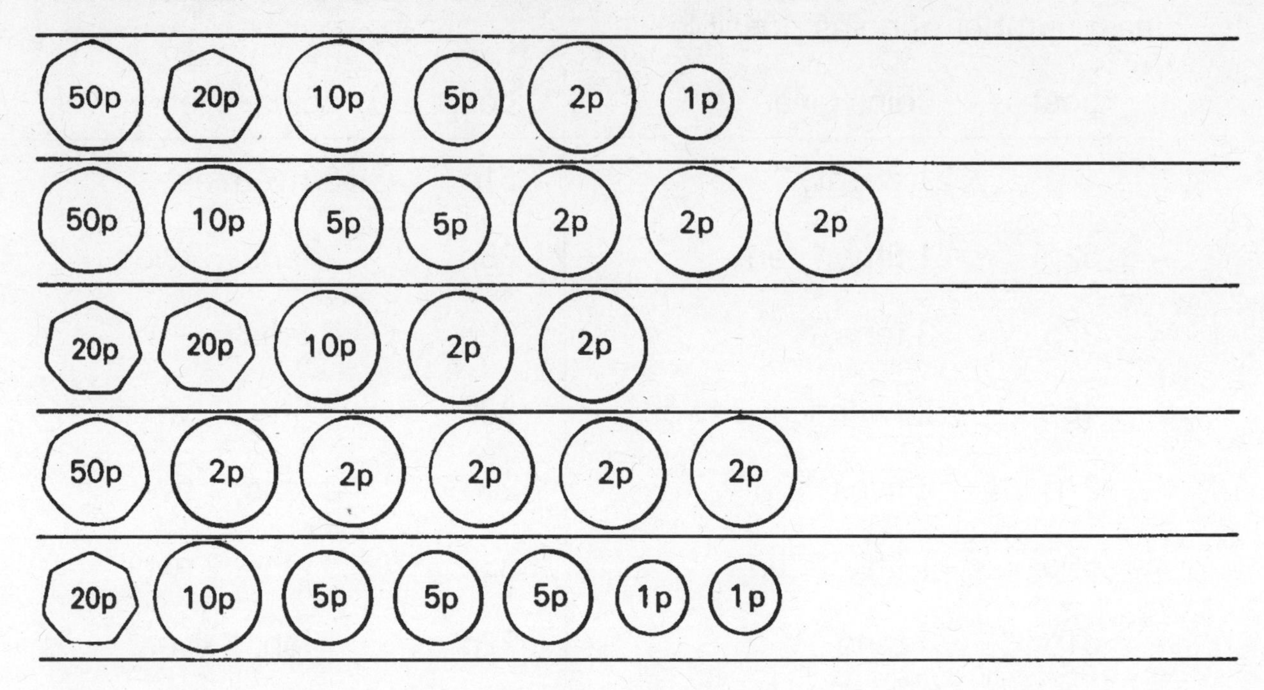

B Which 4 coins make each of these amounts?

18p 15p 22p 30p 16p 5p 57p 75p

8p 35p 80p 64p 19p 32p 12p 9p

C Write the coins you would receive in change from 50p after spending each amount, if you received the least number of coins possible.

43p 37p 29p 46p 22p 28p

33p 10p 38p 27p 11p 35p

D Write the coins you would receive in change from £1·00, after spending each amount, if the least number of coins were given.

76p 42p 35p 67p 24p 14p 87p 93p

E Write **ten** different ways of making 50p with coins.

Write **ten** different ways of making £1·00 with coins.

Money – addition and subtraction

A

27p	46p	18p	34p	38p	24p	46p
+8p	+7p	+26p	+17p	+26p	+58p	+27p

23p	44p	32p	53p	67p	76p	53p
+38p	+26p	+19p	+28p	+27p	+15p	+18p

23p	24p	23p	47p	36p	52p	28p
42p	57p	25p	8p	25p	36p	42p
+16p	+12p	+38p	+32p	+19p	+8p	+11p

26p	18p	42p	42p	52p	34p	27p
17p	34p	23p	7p	18p	22p	45p
+14p	+26p	+18p	+41p	+16p	+33p	+16p

B

26p	35p	69p	72p	88p	77p	89p
–4p	–2p	–9p	–1p	–6p	–9p	–5p

35p	82p	58p	82p	58p	63p	78p
–7p	–6p	–7p	–5p	–9p	–5p	–8p

47p	63p	71p	80p	96p	32p	54p
–29p	–22p	–38p	–47p	–55p	–27p	–29p

47p	63p	52p	84p	96p	50p	92p
–28p	–37p	–46p	–53p	–47p	–27p	–34p

C What is the total of 47p, 20p, 31p?

Find the difference between 52p and 37p.

How much is 62p less than 85p?

Find the sum of 32p, 25p, and 41p.

16p + 27p + 8p + 7p = ☐ p

83p minus 47p

Money – multiplication and division

A

7p × 2 = ☐ p	4p × 6 = ☐ p	6p × 9 = ☐ p
8p × 10 = ☐ p	3p × 12 = ☐ p	7p × 8 = ☐ p
5p × 11 = ☐ p	9p × 3 = ☐ p	5p × 4 = ☐ p
9p × 6 = ☐ p	11p × 8 = ☐ p	12p × 7 = ☐ p
3p × 9 = ☐ p	8p × 12 = ☐ p	
4p × 7 = ☐ p	10p × 8 = ☐ p	7p × 9 = ☐ p
9p × 4 = ☐ p	2p × 12 = ☐ p	8p × 5 = ☐ p
3p × 11 = ☐ p	5p × 3 = ☐ p	12p × 2 = ☐ p
11p × 7 = ☐ p	4p × 9 = ☐ p	6p × 6 = ☐ p

14p ×5	22p ×4	19p ×3	18p ×5	13p ×7	23p ×4	17p ×3
17p ×2	12p ×7	19p ×4	23p ×3	16p ×4	34p ×2	

B

4⟌96p	8⟌72p	3⟌81p	9⟌63p	6⟌84p	2⟌76p	5⟌95p
2⟌42p	4⟌76p	8⟌96p	6⟌78p	2⟌84p	6⟌72p	4⟌72p
3⟌48p	5⟌80p	7⟌84p	9⟌99p	5⟌65p	3⟌81p	7⟌91p

C

1. Find the cost of 6 pens at 12p each.
2. Four cakes cost 92p. How much for each cake?
3. 18p + 18p + 18p + 18p. Find the total by multiplication.
4. Six children share 84p. How much each?
5. One lolly costs 13p. What is the cost of **a** 4 lollies? **b** 6 lollies? **c** 3 lollies? **d** 7 lollies **e** 2 lollies?
6. Share 1 fifty, 2 tens and 1 two equally among Emily, Jack, Charlie and Chloe.

Pounds and pence

100p = £1·00

Complete:

A 200p = £ 500p = £ 700p = £ 600p = £

900p = £ 300p = £ 800p = £ 400p = £

B 104p = £ 305p = £ 143p = £

236p = £ 624p = £ 379p = £

476p = £ 217p = £ 914p = £

807p = £ 763p = £ 207p = £

C £3·45 = £ + p £6·17 = £ + p

£7·06 = £ + p £9·73 = £ + p

£12·14 = £ + p £15·48 = £ + p

£2·72 = £ + p £8·05 = £ + p

D £4·17 = p £8·12 = p £9·08 = p

£8·36 = p £6·18 = p £3·12 = p

£2·04 = p £5·42 = p £5·23 = p

£6·25 = p £4·38 = p £4·77 = p

E £1·16 = £1 + 1 ten + 6 ones

£2·72 = £ + tens + ones

£6·42 = £ + tens + ones

£9·04 = £ + tens + ones

£7·72 = £ + tens + ones

£8·96 = £ + tens + ones

F How many 1p in:

£2·70? £4·36? £7·53? £9·47? £10·36?

How many 10p in:

£9·20? £3·40? £2·50? £6·70? £8·30?

Pounds and pence

A Write as pence:

£0·17 £0·27 £0·36 £0·76 £0·89 £0·05

B Write as pounds:

26p 7p 34p 8p 52p 5p

C Write in figures, using the £ sign:

four pounds twenty-seven	seventy-six pence
one pound eighteen	six pence
two pounds four	ninety-six pence

D Write in words:

£7·45	£0·72	£14·04	£3·62
£0·12	£5·32	£9·26	£0·07

E Write these amounts in order with the largest first.

a	£4·00	414p	47p	£4·36	£0·44
b	505p	£5·50	£5·27	59p	£0·61
c	£2·02	200p	22p	£22·00	£2·20

F Add these amounts of money. Write you answers in pounds.

46p +63p	54p +67p	82p +34p	29p +63p	57p +72p	83p +27p
52p +18p	87p +34p	62p +53p	40p +63p	38p +61p	24p +92p

Addition and subtraction – £ and p

A

£0·17	£0·48	£0·63	£0·53	£0·56
+£0·43	+£0·37	+£0·26	+£0·29	+£0·32

£0·74	£0·84	£0·76	£0·67	£0·29
+£0·32	+£0·42	+£0·57	+£0·63	+£0·94

£4·07	£5·18	£6·42	£8·02	£5·34
+£2·66	+£3·82	+£2·97	+£1·97	+£7·69

£2·02	£2·11	£1·48	£6·16	£0·85
£3·54	£4·30	£5·86	£3·58	£7·47
+£2·90	+£0·63	+£3·29	+£0·47	+£9·70

B

£0·42	£0·30	£0·62	£0·70	£0·83
−£0·37	−£0·17	−£0·34	−£0·53	−£0·54

£1·63	£1·29	£1·53	£1·27	£1·03
−£0·47	−£0·69	−£0·67	−£0·64	−£0·46

£5·27	£6·34	£8·19	£5·26	£7·32
−£3·19	−£2·83	−£5·19	−£4·47	−£6·29

C Find the sum of £4·26, £0·47 and £2·32.

Find the difference between £6·07 and £3·29.

£12·46 add £3·52 add £0·47 add £0·20.

Six pounds ninety-two subtract eighty-four pence.

How much is £2·82 less than £10·06?

Find the total of £0·27, £6·29, £8·99.

Multiplication – £ and p

A

£0·18 ×5	£0·24 ×4	£0·32 ×3	£0·47 ×2	£0·15 ×6
£0·19 ×5	£0·14 ×6	£0·13 ×7	£0·09 ×9	£0·08 ×11
0·53 ×8	£0·76 ×6	£0·38 ×10	£0·26 ×12	£0·64 ×9
£0·32 ×9	£0·67 ×5	£0·93 ×7	£0·26 ×11	£0·32 ×12
£1·43 ×12	£2·76 ×10	£3·07 ×8	£4·32 ×9	£1·59 ×6
£2·07 ×3	£1·24 ×6	£3·18 ×7	£6·43 ×2	£4·99 ×9

B £4·22 × 3   £ 3·34 × 7

£3·17 × 6   £12·16 × 9

£2·18 × 9   £ 8·43 × 12

C Find the product of:

£4·23 and 6   £2·20 and 8

D Multiply:

£12·06 by 8   £9·26 by 3

E Solve by multiplication:

£2·71 + £2·71 + £2·71   £0·27 + £0·27 + £0·27

Division – £ and p

A

$7\overline{)\pounds 0{\cdot}42}$	$4\overline{)\pounds 0{\cdot}96}$	$6\overline{)\pounds 0{\cdot}84}$	$5\overline{)\pounds 0{\cdot}85}$	$3\overline{)\pounds 0{\cdot}87}$
$6\overline{)\pounds 0{\cdot}42}$	$8\overline{)\pounds 0{\cdot}96}$	$4\overline{)\pounds 0{\cdot}40}$	$2\overline{)\pounds 0{\cdot}92}$	$10\overline{)\pounds 0{\cdot}60}$
$7\overline{)\pounds 0{\cdot}84}$	$5\overline{)\pounds 0{\cdot}65}$	$9\overline{)\pounds 0{\cdot}81}$	$11\overline{)\pounds 0{\cdot}77}$	$3\overline{)\pounds 0{\cdot}78}$
$12\overline{)\pounds 1{\cdot}92}$	$8\overline{)\pounds 1{\cdot}84}$	$3\overline{)\pounds 1{\cdot}41}$	$6\overline{)\pounds 0{\cdot}72}$	$9\overline{)\pounds 1{\cdot}98}$
$6\overline{)\pounds 1{\cdot}56}$	$10\overline{)\pounds 1{\cdot}70}$	$12\overline{)\pounds 1{\cdot}80}$	$8\overline{)\pounds 1{\cdot}44}$	$4\overline{)\pounds 1{\cdot}28}$
$3\overline{)\pounds 1{\cdot}32}$	$11\overline{)\pounds 1{\cdot}87}$	$5\overline{)\pounds 1{\cdot}90}$	$9\overline{)\pounds 1{\cdot}98}$	$7\overline{)\pounds 1{\cdot}75}$
$12\overline{)\pounds 52{\cdot}32}$	$7\overline{)\pounds 12{\cdot}67}$	$6\overline{)\pounds 8{\cdot}34}$	$11\overline{)\pounds 6{\cdot}71}$	$9\overline{)\pounds 10{\cdot}80}$
$8\overline{)\pounds 21{\cdot}60}$	$6\overline{)\pounds 19{\cdot}02}$	$2\overline{)\pounds 11{\cdot}24}$	$12\overline{)\pounds 47{\cdot}16}$	$4\overline{)\pounds 8{\cdot}76}$
$7\overline{)\pounds 34{\cdot}30}$	$11\overline{)\pounds 42{\cdot}13}$	$5\overline{)\pounds 14{\cdot}35}$	$9\overline{)\pounds 28{\cdot}44}$	$3\overline{)\pounds 8{\cdot}43}$

B

£11·52 ÷ 12	£6·12 ÷ 9
£55·15 ÷ 5	£65·87 ÷ 7
£2·34 ÷ 6	£12·06 ÷ 3
£3·04 ÷ 4	£6·02 ÷ 2
£73·70 ÷ 11	£48·32 ÷ 8

C

Share: £33·04 by 7
£8·28 by 9
£60·15 by 5

Divide: £27·84 by 12
£3·74 by 11
£74·97 by 7

Buying

1 Find the cost of a kite and a large ball.

2 What is the total cost of an aeroplane, roller skates and a chess set?

3 What would be the change from £1·00 after buying:

a a car? **b** a large ball? **c** crayons?

4 How much dearer is the train set than the chess set?

5 What is the cost of 3 pairs of roller skates?

6 Find the cost of:

a 4 dolls **b** 3 aeroplanes **c** 4 cars **d** 5 kites **e** 6 colouring books.

7 What is the difference between the cost of an aeroplane and a sewing set?

8 What change would I receive from £5·00 if I bought a book, crayons and a car?

9 How many kites could I buy for £4·00 and what would my change be?

10 Which would cost the most, 3 aeroplanes or 4 dolls? How much would the difference in price be?

Capacity

1 litre = 1000 ml

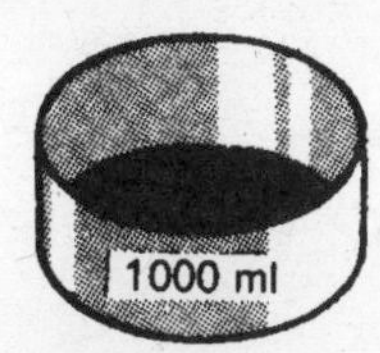

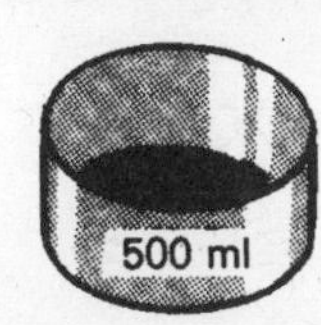

A How many millilitres in:

1	1 l?	2 l?	4 l?	6 l?
	9 l?	7 l?	3 l?	5 l?
	$\frac{1}{2}$ l?	$\frac{1}{4}$ l?	$1\frac{1}{2}$ l?	$1\frac{1}{4}$ l?
	$\frac{3}{4}$ l?	$2\frac{1}{4}$ l?	$2\frac{3}{4}$ l?	$1\frac{3}{4}$ l?

$$\frac{1}{10}\ \text{l} = 0{\cdot}1\ \text{l} = 100\ \text{ml}$$

2	0·5 l?	0·4 l?	0·9 l?	0·7 l?
	0·3 l?	0·8 l?	0·2 l?	0·1 l?
	2·4 l?	1·6 l?	3·8 l?	4·5 l?
	6·1 l?	5·3 l?	8·8 l?	2·7 l?

B Write down **six** different ways you could fill the 1 litre container using the 500 ml, 250 ml and 125 ml containers.

C Write down **three** ways you could fill the 500 ml container using the 250 ml and the 125 ml containers.

D

milk

juice

medicine

milk

lemonade

pop

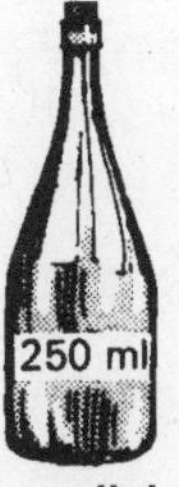

cordial

1. Which two containers together hold $\frac{1}{2}$ litre?
2. How many bottles of cordial make a litre?
3. Which two containers together would fill the 600 ml milk bottle?
4. How much more does the juice bottle hold than the large milk bottle?
5. How many small milk bottles would fill 2 medicine bottles?
6. Which 3 containers together hold the same as the juice bottle?

Length

1 metre = 100 cm

A Change to cm:

2 m 9 m 6 m 4 m 7 m 12 m 16 m

B Change to m:

400 cm 300 cm 500 cm 100 cm 800 cm 1300 cm

C Complete:

$\frac{1}{2}$ m = cm	$\frac{1}{4}$ m = cm	$\frac{3}{4}$ m = cm
$2\frac{1}{2}$ m = cm	$1\frac{1}{4}$ m = cm	$1\frac{3}{4}$ m = cm
$4\frac{1}{2}$ m = cm	$2\frac{1}{4}$ m = cm	$2\frac{3}{4}$ m = cm
$6\frac{1}{2}$ m = cm	$4\frac{1}{4}$ m = cm	$7\frac{3}{4}$ m = cm

D Use fractions:

50 cm = m	25 cm = m	75 cm = m
350 cm = m	525 cm = m	375 cm = m
150 cm = m	325 cm = m	475 cm = m

E

1 m 23 cm = cm	2 m 16 cm = cm
4 m 5 cm = cm	3 m 8 cm = cm
9 m 36 cm = cm	8 m 29 cm = cm

F

225 cm = m cm	518 cm = m cm
312 cm = m cm	654 cm = m cm
806 cm = m cm	931 cm = m cm

321 cm = 3·21 m

G Complete using decimals:

532 cm = m	196 cm = m	92 cm = m
402 cm = m	52 cm = m	307 cm = m
390 cm = m	7 cm = m	1426 cm = m
4·26 m = cm	8·16 m = cm	10·27 m = cm
12·06 m = cm	0·09 m = cm	3·41 m = cm
0·27 m = cm	0·77 m = cm	1·01 m = cm

H Write as decimals:

$\frac{1}{2}$ m $\frac{1}{4}$ m $\frac{3}{4}$ m $\frac{1}{100}$ m $\frac{23}{100}$ m

Length

10 mm = 1 cm

A 50 mm = cm 40 mm = cm 60 mm = cm
30 mm = cm 90 mm = cm 20 mm = cm
130 mm = cm 220 mm = cm 190 mm = cm

B 7 cm = mm 8 cm = mm 9 cm = mm
6 cm = mm 5 cm = mm 3 cm = mm
15 cm = mm 320 cm = mm 530 cm = mm

27 mm = 2 cm 7 mm = 2·7 cm

C Complete:
53 mm = cm mm = cm 7 mm = cm mm = cm
86 mm = cm mm = cm 4 mm = cm mm = cm
131 mm = cm mm = cm 6 mm = cm mm = cm
268 mm = cm mm = cm
320 mm = cm mm = cm

D Change to mm:
4·1 cm 3·7 cm 0·9 cm 0·8 cm 14·2 cm 5·6 cm

1000 mm = 1 metre **7340 mm = 7·340 m**

E Complete:
7000 mm = m 9000 mm = m 3000 mm = m
4500 mm = m 2500 mm = m 8500 mm = m
5250 mm = m 8250 mm = m 1250 mm = m
9750 mm = m 3750 mm = m 4750 mm = m

F Change to mm:
$6\frac{1}{2}$ m $5\frac{1}{2}$ m $3\frac{1}{2}$ m $2\frac{3}{4}$ m $1\frac{3}{4}$ m $4\frac{1}{4}$ m $7\frac{1}{4}$ m $9\frac{1}{4}$ m

G Complete:
6670 mm = m mm 5026 mm = m mm 2340 mm = m mm
8026 mm = m mm 2726 mm = m mm 8791 mm = m mm

H 4 m 135 mm = mm 6 m 296 mm = mm 8 m 196 mm = mm
16 m 176 mm = mm 5 m 17 mm = mm 9 m 5 mm = mm

The kilometre

1 km = 1000 m

A Change to metres:

1 km 275 m	3 km 536 m	4 km 326 m	
8 km 824 m	2 km 56 m	7 km 36 m	
5 km 4 m	6 km 8 m	$4\frac{1}{2}$ km	$3\frac{3}{4}$ km
$2\frac{3}{4}$ km	$9\frac{1}{2}$ km		

B Write as km and m:

3760 m	2430 m	1641 m	
8252 m	5862 m	4026 m	
7054 m	1006 m	3002 m	6008 m

Perimeter

C Measure the sides of these shapes, then write:

a the length **b** the width **c** the perimeter.

A

D

B

C

F

G

E

Area and perimeter

A Find the area of these shapes.

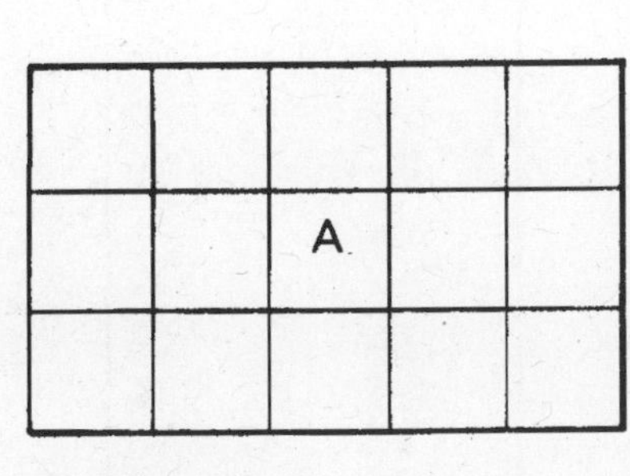

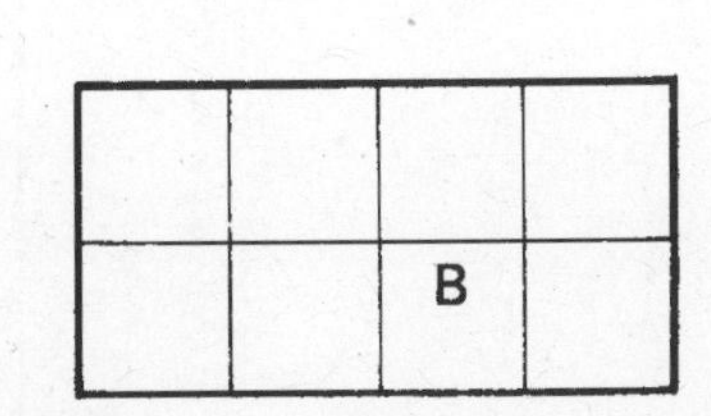

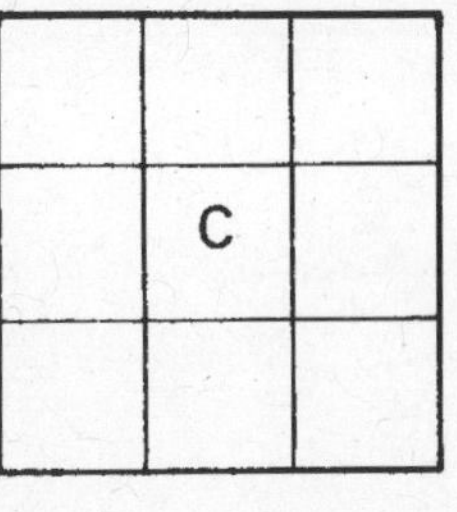

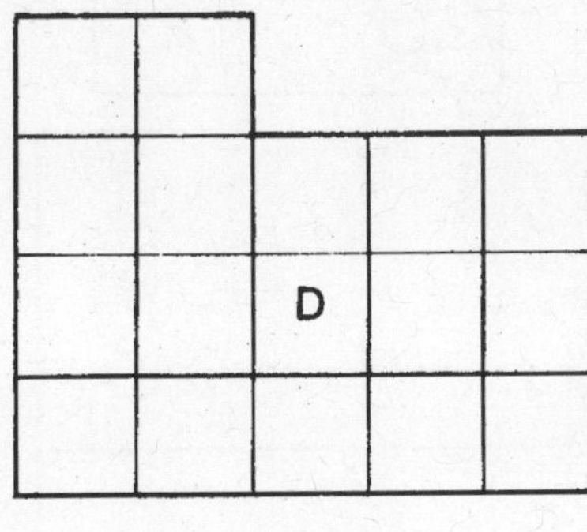

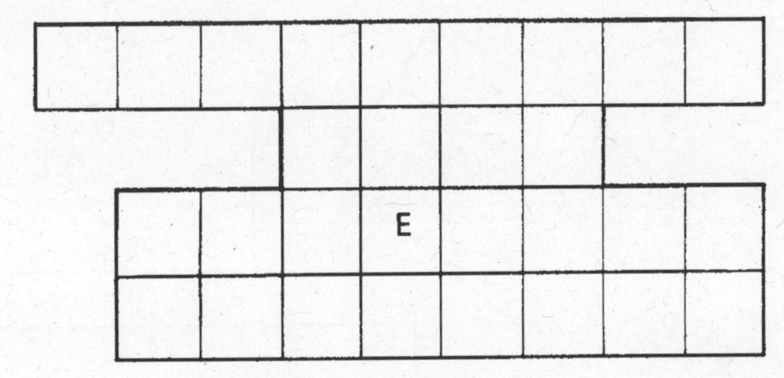

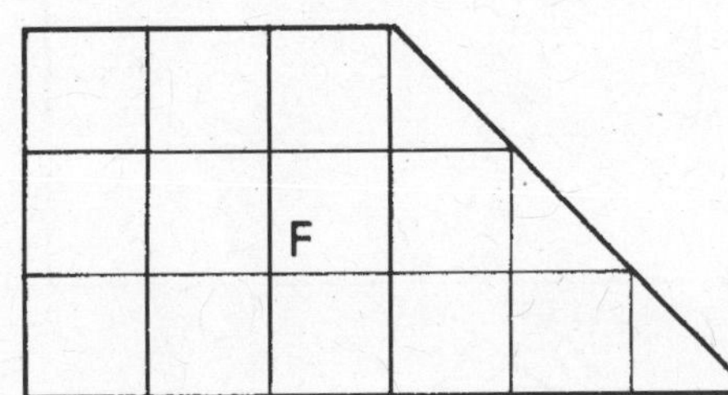

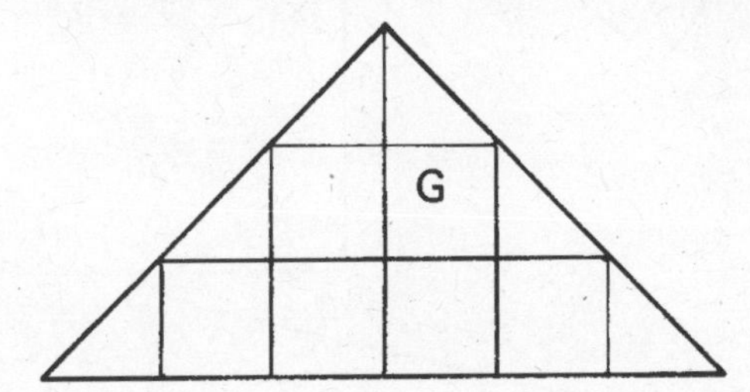

B Find the area of the shapes using length and breadth.

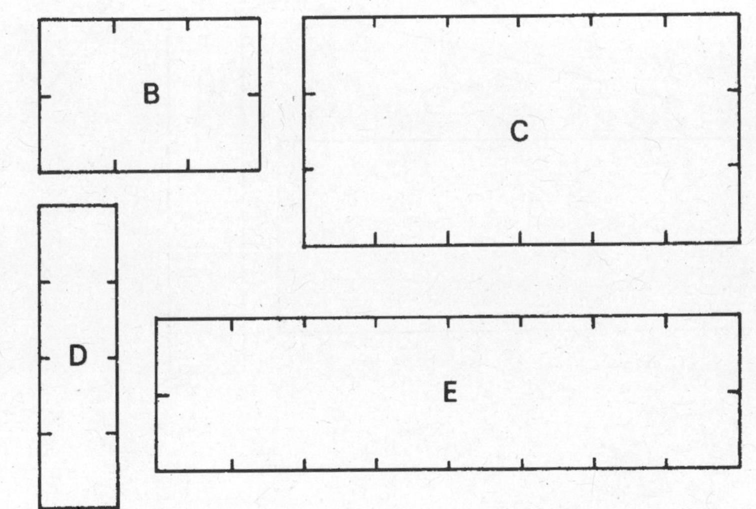

Area and perimeter

Measure the sides of these shapes then find: **a** the perimeter **b** the area.

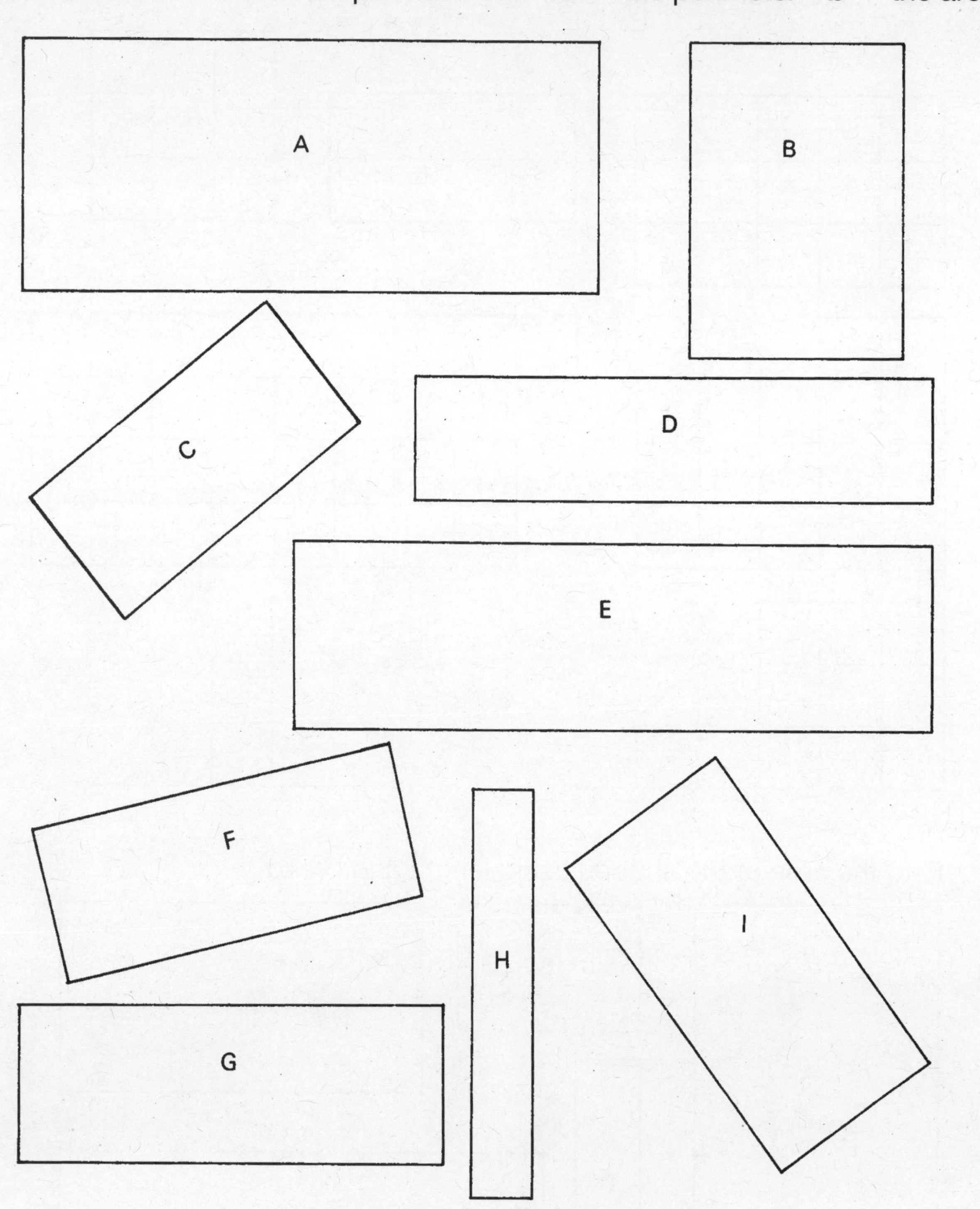

Scale measurement

A Each of these objects has been drawn to a scale of 1 cm to 2 cm. Write the **actual** length of each object.

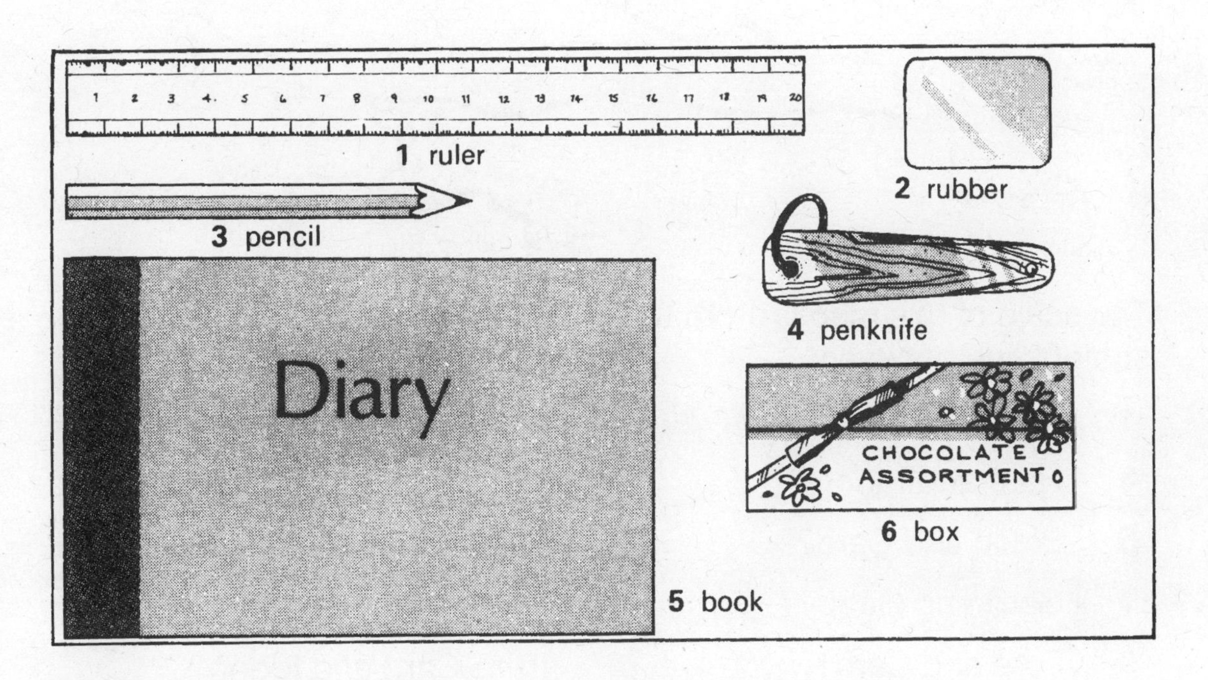

B Using the scale 1 cm to 2 cm draw lines to represent:

20 cm 30 cm 18 cm 14 cm 6 cm 12 cm

C The scale is 1 cm to 5 cm. What length do each of these lines represent?

D The scale is 1 cm to 10 cm. What length of line would you draw to represent these distances?

45 cm 60 cm 55 cm 20 cm 50 cm 25 cm

E The scale is 1 cm to 4 km. How far apart are towns A and B?

A ○———————————————○ B

Scales and maps

A

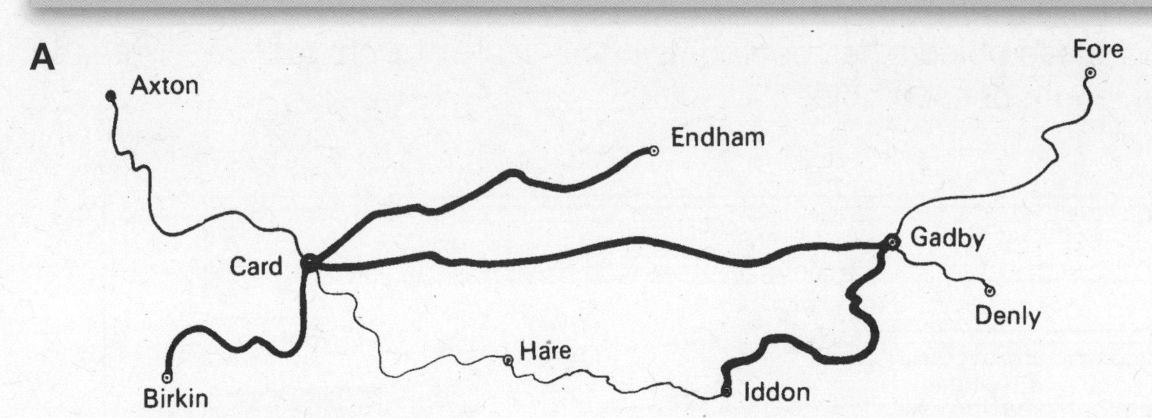

The scale of this map is 1 cm to 20 km. What is the actual distance in a straight line between:

1 Axton and Endham?
2 Card and Endham?
3 Hare and Iddon?
4 Gadby and Denly?
5 Birkin and Card?
6 Iddon and Gadby?
7 Card and Gadby?
8 Gadby and Fore?
9 Fore and Card?
10 Card and Iddon?
11 Axton and Card?
12 Hare and Card?

B Scale 1 cm to 5 km. What distance do these lines represent?

What length of line would represent these distances?

15 km 40 km 25 km 60 km $17\frac{1}{2}$ km $52\frac{1}{2}$ km

C Measure these lines then find the scale which has been used.

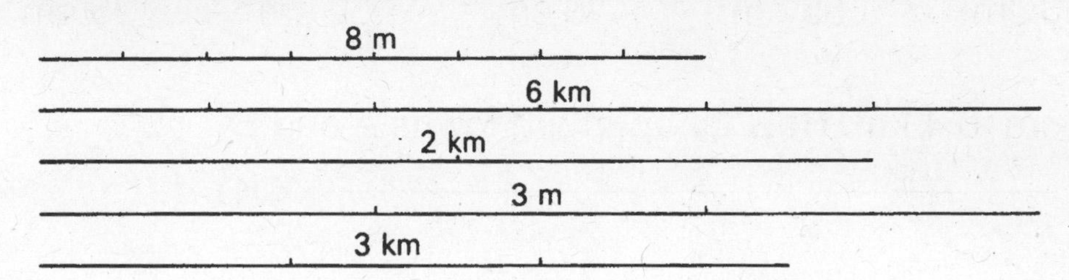

Mass

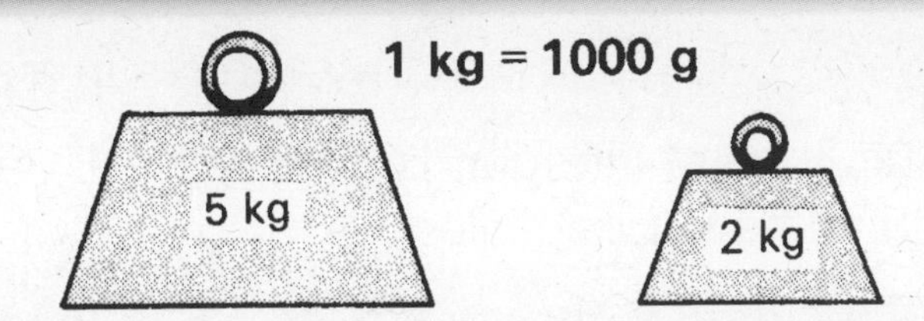

A How many of each of the following weights would it take to balance 5 kg?

500 g 200 g 100 g 50 g 20 g 10 g

B How many grams in:

2 kg? $8\frac{1}{2}$ kg? $6\frac{1}{2}$ kg? 10 kg? $3\frac{1}{2}$ kg? 7 kg?

C How many kilograms in:

4000 g? 6000 g? 8500 g? 2500 g? 5000 g? 3500 g?

D Write the following as kg and g.

4260 g 8120 g 2050 g 3160 g 7005 g

E Write as grams:

2 kg 360 g 3 kg 115 g 5 kg 70 g 4 kg 33 g 6 kg 5 g

F Write as grams:

3·2 kg 4·1 kg 6·3 kg 5·4 kg 8.6 kg

G How much less than $\frac{1}{2}$ kg are each of these weights?

436 g 229 g 347 g 291 g 32 g 67 g

H How much greater than 2 kg are:

3 kg 471 g? 2 kg 126 g? 4021 g?

11 kg 741 g? 8006 g? Give your answers in kg and g.

I Balance these equations.

4 kg 200 g + 800 g = + 3 kg 250 g

700 g + 200 g + 300 g = – 200 g

+ 3 kg + 500 g = 4 kg + 600 g

6 kg – = 2 kg 300 g – 1 kg 800 g

200 g + 800 g + 3 kg 100 g = 4 kg 50 g +

Comparison – other measures

1 Three oranges weigh 500 g. How many oranges in a sack weighing 50 kg?

2 A barrel holds 300 litres of vinegar. How much will be left after 25 bottles each holding 500 ml have been filled?

3 How many 80 cm lengths can be cut from a 5 m length of wood? What length of wood is left?

4 A jar holds 450 ml. How many litres in 12 jars?

5 A box contains 36 blocks of chocolate each weighing 100 g. If the box weighs 50 g when it is empty, what is the total weight? Answer in kg.

6 How many 250 g packs of butter can be weighed from $8\frac{1}{2}$ kg?

7 A rope is cut into 6 equal parts of 75 cm. How long was the length of rope in metres?

8 A car used 4 litres of petrol on a journey of 50 km. How many litres are used on a journey of 725 km?

9 How many 200 ml bottles can be filled from a 50 litre tank of milk?

10 How many 50 g bags of sweets can be weighed from a 5 kg box?

11 A sack holds 50 kg of potatoes. How many 2 kg bags can be made from it?

12 A milk bottle holds 600 ml. If a family has 2 bottles of milk each day, how many litres do they drink in a week?

13 How long are 8 pencils if one pencil is 17 cm long?

14 Four lengths of cloth each $2\frac{1}{4}$ m long are cut from a roll of cloth 20 m long. How many more lengths of $2\frac{1}{4}$ m can be cut from it?

15 A $\frac{1}{4}$ of a fruit cake weighs 425 g. How much does the whole cake weigh? Answer in kg.

16 A boy's stride is 50 cm. How many steps does he take to cover 100 metres?

17 If ten sweets weigh 50 g, what would 210 sweets weigh? Answer in kg.

18 A medicine spoon holds 5 ml. How many doses are there in a 300 ml bottle?

19 12 balls of wool each weighing 25 g are needed to knit a girl's cardigan. How much wool is needed to knit 5 cardigans? Answer in kg.

20 **a** How many ribbons each measuring 12 cm can be cut from a 4 m roll?

b What length of ribbon will be left?

Shopping

Find the cost of the following amounts.

		a	b	c	d
1	1 litre costs 80p	4 litres	6 litres	10 litres	9 litres
2	$\frac{1}{2}$ litre costs 10p	$\frac{1}{4}$ litre	$1\frac{1}{4}$ litres	$4\frac{1}{2}$ litres	6 litres
3	400 ml cost 32p	100 ml	600 ml	700 ml	900 ml
4	600 ml cost 60p	900 ml	400 ml	450 ml	750 ml
5	500 ml cost 16p	$2\frac{1}{2}$ l	3 l	$\frac{1}{4}$ l	$1\frac{1}{4}$ l
6	$\frac{1}{4}$ litre costs 15p	$\frac{3}{4}$ l	$\frac{1}{2}$ l	$4\frac{1}{4}$ l	$2\frac{3}{4}$ l
7	$\frac{1}{2}$ metre costs £1·25	5 m	$8\frac{1}{2}$ m	6 m	$3\frac{1}{2}$ m
8	$\frac{1}{2}$ metre costs 60p	$\frac{1}{4}$ m	$2\frac{1}{4}$ m	$3\frac{1}{2}$ m	5 m
9	1 metre costs 96p	3 m	$2\frac{1}{2}$ m	5 m	7 m
10	$\frac{1}{4}$ metre costs 12p	1 m	$\frac{1}{2}$ m	$2\frac{1}{2}$ m	$3\frac{1}{4}$ m
11	$\frac{3}{4}$ metre costs 33p	$\frac{1}{4}$ m	1 m	$5\frac{1}{2}$ m	$2\frac{1}{4}$ m
12	$\frac{1}{4}$ metre costs 35p	$4\frac{1}{2}$ m	$1\frac{1}{4}$ m	3 m	6 m
13	$\frac{1}{2}$ kg costs 35p	$2\frac{1}{2}$ kg	$1\frac{1}{2}$ kg	$3\frac{1}{2}$ kg	5 kg
14	$\frac{1}{2}$ kg costs 16p	$\frac{1}{4}$ kg	$5\frac{1}{2}$ kg	$6\frac{1}{4}$ kg	7 kg
15	500 g cost 25p	100 g	700 g	900 g	300 g
16	250 g cost 7p	1 kg	$1\frac{1}{2}$ kg	$2\frac{1}{4}$ kg	4 kg
17	200 g cost 12p	400 g	500 g	1 kg	$1\frac{1}{4}$ kg
18	100 g cost 4p	1 kg	400 g	700 g	$5\frac{1}{2}$ kg

Time

A Write in words, the times shown on these clocks.

B Write in words, the times shown on these clocks in two ways.

C Draw clocks to show these times.

ten past six	five to nine	$\frac{1}{4}$ past eight
35 minutes past two	$\frac{1}{2}$ past one	40 minutes past three
$\frac{1}{4}$ to twelve	twenty past five	ten to four
50 minutes past seven	twenty-five past eleven	five past ten
two o'clock	20 minutes to three	25 minutes to six

Time

A Write these times using figures only.

10 minutes past 6 | quarter to nine | 10 minutes to 4

quarter past eight | 20 minutes to 5 | 5 minutes to 10

B Write the times shown on these clocks using figures only.

C Write these times using **am** or **pm**.

morning times

afternoon and evening times

D Write these times in figures using **am** or **pm**.

25 minutes to one in the afternoon

20 minutes past six in the afternoon

10 minutes to four in the morning

quarter to eight in the evening

25 minutes past twelve in the morning

5 minutes to nine in the evening

quarter past two in the morning

10 minutes past eleven in the evening

5 minutes past three in the afternoon

half past five in the morning

Time

A

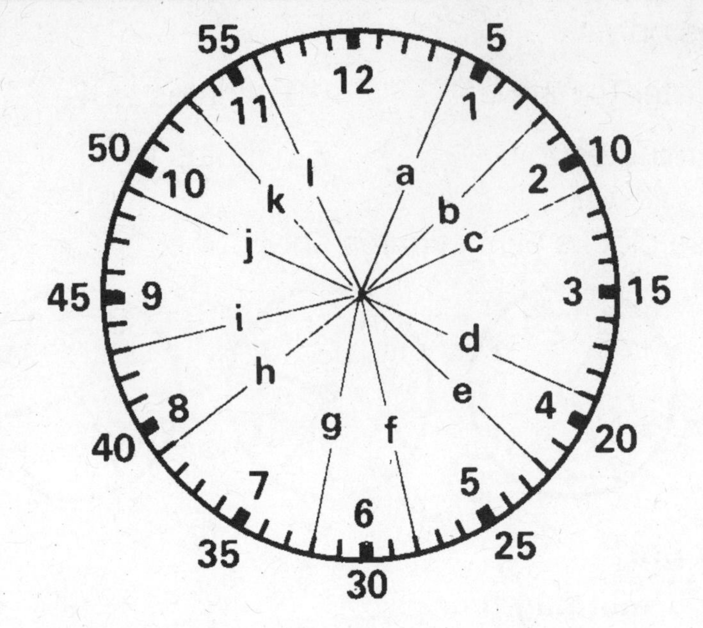

Write the number of minutes shown by each of the pointers.

a	**b**	**c**
d	**e**	**f**
g	**h**	**i**
j	**k**	**l**

B Write the time shown on each of these clocks in figures.

   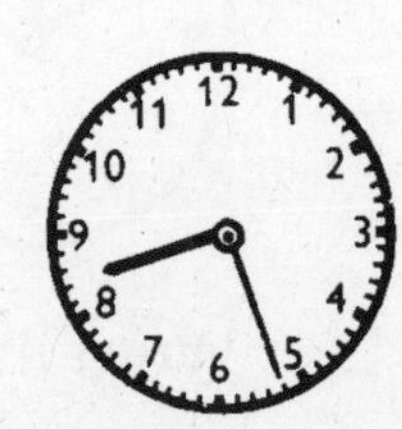

C Write the times shown on each clock in figures using **am** or **pm**.

morning

evening

afternoon

morning

Adjusting times

A Each clock is 10 minutes slow. What is the correct time?

B Each clock is 10 minutes fast. What is the correct time?

C How many minutes must pass for these clocks to show 10 o'clock?

D

A

B

C

D

How many minutes have passed between clocks:

A and B? B and C?

A and C? B and D?

A and D? C and D?

Time – problems

A Here are the times of the bus between Porton and Fartown. Answer the questions.

town	time
Porton	8:15
Benford	8:39
Markham	8:57
Southam	9:20
Fartown	9:55

1 How long does the journey from Porton to Fartown take?

2 How long from Porton to Benford?

3 How long from Porton to Markham?

4 How long from Porton to Southam?

5 Between which two towns does the journey take the least time?

6 Between which two towns does the journey take the most time?

B Find how many hours and minutes between these times.

8:00 and 10:30   0:55 and 1:25

7:36 and 7:55   3:29 and 3:51

2:29 and 3:42   11:23 and 12:06

3:31 and 5:26   1:42 and 4:03

C How many minutes in:

$1\frac{1}{2}$ h? $1\frac{1}{4}$ h? $2\frac{1}{2}$ h? $4\frac{1}{2}$ h?

$1\frac{3}{4}$ h? $2\frac{1}{4}$ h? $2\frac{3}{4}$ h?

D A TV programme starts at 5:40 pm and lasts for 45 min. What time does it end?

E Oliver's watch says 2:55. If it is a quarter of an hour slow, what is the correct time?

F School starts at 9:00 am Jake leaves home at 8:55 am and is 3 minutes late. How long did he take to reach school?

G Lunchtime at school lasts from 12:00 pm to 1:15 pm.

1 How many minutes is this?

2 If afternoon school lasts $2\frac{1}{2}$ h, what time does school close?

The calendar

January

Su	M	Tu	W	Th	F	S
	1	2	3	4	5	6
7	8	9	10	11	12	13
14	15	16	17	18	19	20
21	22	23	24	25	26	27
28	29	30	31			

February

Su	M	Tu	W	Th	F	S
				1	2	3
4	5	6	7	8	9	10
11	12	13	14	15	16	17
18	19	20	21	22	23	24
25	26	27	28			

March

Su	M	Tu	W	Th	F	S
				1	2	3
4	5	6	7	8	9	10
11	12	13	14	15	16	17
18	19	20	21	22	23	24
25	26	27	28	29	30	31

April

Su	M	Tu	W	Th	F	S
1	2	3	4	5	6	7
8	9	10	11	12	13	14
15	16	17	18	19	20	21
22	23	24	25	26	27	28
29	30					

May

Su	M	Tu	W	Th	F	S
		1	2	3	4	5
6	7	8	9	10	11	12
13	14	15	16	17	18	19
20	21	22	23	24	25	26
27	28	29	30	31		

June

Su	M	Tu	W	Th	F	S
					1	2
3	4	5	6	7	8	9
10	11	12	13	14	15	16
17	18	19	20	21	22	23
24	25	26	27	28	29	30

July

Su	M	Tu	W	Th	F	S
1	2	3	4	5	6	7
8	9	10	11	12	13	14
15	16	17	18	19	20	21
22	23	24	25	26	27	28
29	30	31				

August

Su	M	Tu	W	Th	F	S
			1	2	3	4
5	6	7	8	9	10	11
12	13	14	15	16	17	18
19	20	21	22	23	24	25
26	27	28	29	30	31	

September

Su	M	Tu	W	Th	F	S
						1
2	3	4	5	6	7	8
9	10	11	12	13	14	15
16	17	18	19	20	21	22
23	24	25	26	27	28	29
30						

October

Su	M	Tu	W	Th	F	S
	1	2	3	4	5	6
7	8	9	10	11	12	13
14	15	16	17	18	19	20
21	22	23	24	25	26	27
28	29	30	31			

November

Su	M	Tu	W	Th	F	S
				1	2	3
4	5	6	7	8	9	10
11	12	13	14	15	16	17
18	19	20	21	22	23	24
25	26	27	28	29	30	

December

Su	M	Tu	W	Th	F	S
						1
2	3	4	5	6	7	8
9	10	11	12	13	14	15
16	17	18	19	20	21	22
23	24	25	26	27	28	29
30	31					

1. Is this year a leap year? How do you know?
2. Which 4 months have 5 Fridays in them?
3. Which 2 months begin on Sunday?
4. Name the 2 months which end on Saturday.
5. Olivia is six months younger than Charlie. Charlie's birthday is in December. When is Olivia's birthday?
6. Olivia's school went on an outing on the third Wednesday of the sixth month. What date was this?
7. Discos were held on the second Saturday of each month. Give the dates of the 12 discos this year.
8. Give the day on which each month with 31 days ends.
9. Give the day on which each month with 30 days begins.
10. On what day was New Year's Day?
11. On what day will New Year's Day next year be?
12. On what day was Christmas Day?

Graphs

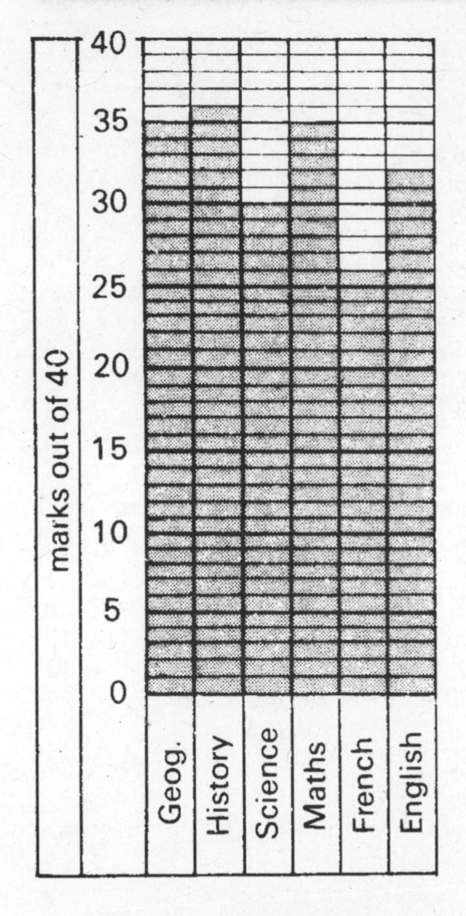

A The graph shows how many marks Harry gained in his examinations. Copy and complete this table.

Subject	Geog.	History	Science	Maths	French	English
mark						

1. In which subject did he gain highest marks?
2. Which subject was his worst?
3. In which subjects did he gain the same marks?
4. In which subject did he gain $\frac{3}{4}$ marks?
5. What was the difference between his highest mark and his lowest mark?
6. What were the total possible marks?
7. What was Harry's total mark?
8. How many marks did he lose altogether?

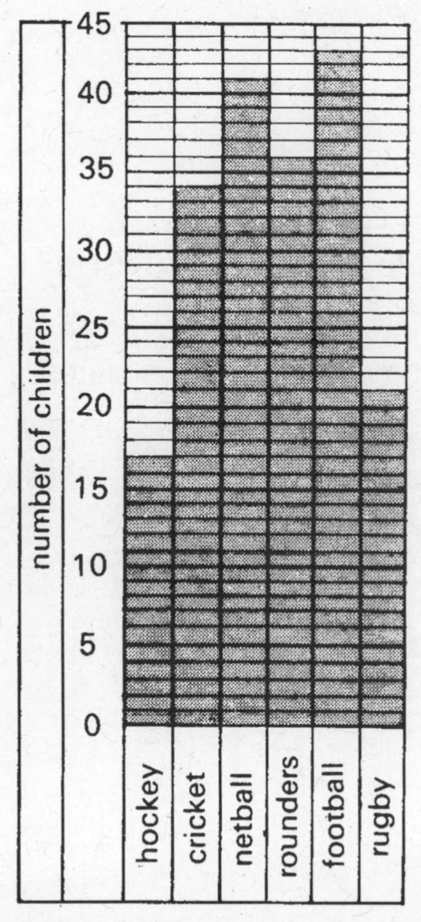

B **Brownhill school**

Each child in the school chose their favourite sports game.

Hockey, netball and rounders were for girls.

Rugby, cricket and football were for boys.

How many voted for each game?

hockey	cricket	netball	rounders	football	rugby

1. How many boys voted?
2. How many girls voted?
3. How many children voted altogether?
4. There are 200 children in the school. How many were absent?
5. What was the boys' favourite game?
6. What was the girls' favourite game?
7. How many chose rugby or football?
8. How many chose hockey or rounders?
9. Which game received the least votes?
10. How many more boys voted than girls?

Graphs

A

0 5 10 15

Heaton
Warcombe
Corm
Birford
Forham
Wessex
Imford
Frame

Schools' football

This is a record of the football matches played between eight school teams, showing the matches won.

1. Which team won the most matches?
2. Which team won the fewest matches?
3. Which teams won the same number of matches?
4. Which two teams together won the same number of matches as Warcombe?
5. How many more matches did Heaton win than Birford?
6. Which two teams together won as many matches as Imford and Warcombe?

B **High school attendance – Monday**

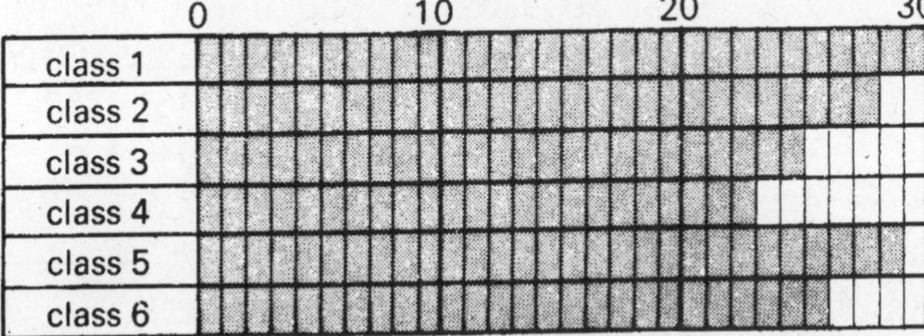

Each class has 30 children.

1. How many children are there in the school?
2. How many children were absent on Monday?
3. Which class had the most absences?
4. Which class had full attendance?

C Draw your own graphs for these tables, done by a group of children about their class.

1 Pets

horse	rabbit	dog	cat	fish	hamster
2	9	15	6	5	3

2 Drinks

orange	coke	lemon	milk	tea
10	18	5	4	3

Answers

Page 2 Addition
A 78, 79, 89, 99, 89, 98, 98
B 989, 799, 989, 999, 899, 988, 879
C 98, 87, 109, 80, 102, 96, 88
D 894, 986, 993, 798, 898, 796, 999
E 176, 199, 189, 208, 219, 179, 169
F 947, 956, 968, 768, 979, 959, 828
G 131, 143, 155, 154, 137, 145, 193
H 652, 769, 785, 922, 802, 856, 888

Page 3 Addition
A 1697, 1399, 1599, 1478, 1979, 1589
B 1458, 1758, 1799, 1403, 1744, 1287
C 1619, 1619, 1958, 1166, 1714, 1448
D 2439, 2764, 2251, 2193, 2243, 2244
E 2035, 1983, 2128, 2204, 2257, 2105
F 2072, 2364, 2637, 2073, 2533, 2201
G 4489, 4749, 9058, 9749, 7648
H 8311, 8929, 9556, 8663, 9697

Page 4 Subtraction
A 542, 517, 523, 816, 121, 251
B 301, 601, 101, 302, 401, 401
C 200, 300, 100, 400, 100, 400
D 313, 422, 504, 515, 423, 311
E 171, 344, 33, 151, 112, 120
F 209, 416, 212, 419, 318, 408
G 193, 153, 253, 494, 193, 273
H 95, 288, 279, 165, 368, 98
I 127, 214, 8, 33, 211, 41
J 119, 269, 159, 238, 178, 158
K 8, 25, 9, 25; 16, 37, 7, 23;
8, 44, 13, 20

Page 5 Subtraction
A 524, 624, 121, 243, 313, 711
B 736, 913, 644, 824, 947, 724
C 883, 891, 883, 775, 274, 671
D 899, 887, 669, 776, 775, 456
E 649, 529, 476, 128, 266, 569
F 565, 733, 125, 508, 186, 542
G 1556, 1618, 4142, 606, 1322, 3293
H 559, 1234, 3671, 5848, 2957, 4728
I 1468, 3274, 4153, 346, 3641, 371
J 1266, 1489, 5077, 1375, 5579, 893
K 47, 43, 18, 29; 35, 9, 27, 9;
25, 15, 28, 37

Page 6 Notation
A 8 thousands, 3 tens, 4 hundreds,
6 units, 2 hundreds;
2 tens, 4 thousands, 6 units, 5 hundreds,
1 ten
B 620, 60; 850, 85; 115, 90;
710, 200; 6054, 49; 3006, 906
C four thousand and seven,
seven hundred and ninety-three,
one hundred and fifteen,
five hundred and five,
nine thousand six hundred and twenty,
twenty;
thirty, four hundred and two,
five hundred and thirty,
one hundred and seven,
three thousand and seventeen,
nine hundred and ninety-nine
D 100, 410, 421, 1200; 119, 142, 59, 1229;
203, 2300, 196, 1003; 93, 164, 3067, 696;
1041, 2404, 1821, 374; 472, 2241, 4731,
1900

Page 7 Addition and subtraction
A 3794, 1917, 2205, 3746, 7838;
9161, 2819, 3279, 6388, 4471
B 3902, 712; 1055, 325; 8365, 2017;
662, 3202; 334, 5229; 3645, 5772
C 4587, 3724, 4776, 1030, 7949
D 584, 205, 930, 3303, 378

Page 8 Multiplication
A 848, 336, 488, 999, 966, 848
B 510, 642, 832, 618, 618, 636
C 681, 692, 595, 690, 585, 496
D 780, 906, 688, 955, 728, 849
E 582, 702, 996, 945, 876, 534
F 2550, 1850, 1172, 2238, 1196, 3894
G 7355, 7086, 6500, 6378, 3098
H 5444, 7015, 4694, 5820;
4776, 5395, 6998, 5994
I 1380, 380; 5096, 572; 2376, 588
J 615, 1400, 576

Page 9 Multiplication by 7
A 21, 14, 28, 42, 63; 84, 35, 56, 21, 14;
7, 70, 49, 42, 35; 0, 28, 77
B 27, 19, 31, 39; 65, 54, 44, 61;
24, 45, 60, 20; 33, 53, 37, 69;
25, 62, 67, 34; 16, 38, 52

C 2982, 2772, 3199, 1309, 4263, 1036;
2268, 4445, 2086, 3073, 1239, 1456;
8645, 9723, 8232, 7476, 9821
D 3311, 4403, 5656, 6503;
8001, 7602, 8435, 7679
E 3479, 3563, 609, 2653, 672, 2170

Page 10 Multiplication by 8
A 2, 8, 3, 6, 12; 0, 4, 7, 5, 11;
8, 3, 6, 4, 10; 2, 9, 1, 7, 5
B 70, 50, 23, 30, 10; 63, 51, 12, 62, 71;
53, 14, 60, 31, 52; 21, 15, 22, 54, 13;
20, 55, 61, 11
C 2616, 3272, 4944, 4160, 7896, 3448;
4848, 7176, 4256, 1360, 4376, 6584;
8784, 9672, 8560, 9320, 9144
D 3040, 5400, 7912, 1648;
9192, 8552, 8048, 8992
E 3176, 8216, 5832;
3512, 8200, 7432, 5040, 6688, 8072, 4848

Page 11 Multiplication by 9
A 9, 36, 54, 63, 0; 45, 72, 36, 108, 81;
63, 27, 99, 18, 18; 0, 45, 72, 90, 27; 54
B 11, 43, 42, 31, 90; 15, 71, 70, 35, 81;
36, 25, 33, 12, 61; 44, 34, 54, 80, 32;
14, 24, 18, 23, 41; 50, 72, 22, 17, 51;
53, 62, 27, 60, 63; 26, 16, 40, 20, 52;
30, 13, 45, 10, 21
C 3843, 6147, 8451, 1413, 6372, 4734;
5679, 6840, 8181, 3474, 7425, 3879;
9684, 9963, 9315, 9972
D

4**6**	8**0**9	253	**10**4**5**
×9	×9	×9	×9
414	7281	2**2**7**7**	**9**405

Page 12 Multiplication by 11
A 77, 44, 33, 11, 99; 88, 22, 66, 55, 0;
8, 4, 5, 1, 6; 3, 7, 0, 9, 2
B 81, 96, 51, 82, 72; 80, 42, 83, 97, 64;
73, 94, 84, 61, 108; 95, 107, 40, 75, 31;
60, 30, 86, 74, 62; 106, 63, 71, 104, 91;
41, 85, 53, 105, 52; 101, 92, 102, 93, 90;
100, 20, 70, 50, 103
C 2585, 4444, 7458, 1892, 3157, 3916;
1540, 9966, 9075, 8052, 6864, 4202
D 4037, 5577, 1540

Page 13 Multiplication by 12
A 48, 108, 60, 36, 12; 0, 24, 96, 84, 72;
1, 5, 7, 3, 9; 4, 6, 2, 0, 8
B 41, 21, 111, 80, 42; 112, 92, 31, 110, 53;
100, 51, 30, 50, 102; 101, 52, 32, 104, 55;
116, 20, 115, 81, 40; 93, 56, 44, 33, 114;
117, 105, 90, 54, 103; 45, 57, 91, 113, 43
C 8712, 5400, 4500, 1980, 3240, 9624
D 5748, 8748, 3420, 7812;
1956, 5460, 4800, 4728; 9852, 7992
E 5160, 7632, 6252;
4632, 4152, 4908, 9768, 8484

Page 14 Multiplication
A 235, 392, 356, 96, 455, 420, 136;
665, 387, 440, 960, 351, 737, 600;
1376, 2094, 5874, 1802, 6096, 2856;
808, 5148, 2786, 4487, 5520, 3655;
8254, 9144, 7360, 6924, 7553;
9344, 9306, 7042, 7082, 7035
B 464, 1708, 4108, 3490
C 3195, 8608, 5556, 8188
D 9018, 582, 1240, 4988, 7854, 8792, 7084

Page 15 Division
A 333, 333, 222, 111, 111, 222;
92, 51, 61, 51, 31, 71;
172, 192, 151, 161, 161, 181;
191 r 2, 181 r 2, 191 r 2, 141 r 3, 131 r 2,
131 r 2;
213, 119, 119, 114, 327, 239;
287, 162, 156, 486, 144, 194;
108, 209, 109, 309, 109, 409;
112 r 3, 289 r 1, 149 r 3, 254 r 1, 243 r 3,
275 r 1;
170, 150, 140, 190, 160, 130;
330, 210, 230, 110, 130, 410;
200, 200, 200, 300, 200, 400;
176 r 3, 141 r 3, 234 r 2, 150 r 4, 134 r 1,
161 r 2;
197 r 2, 94 r 3, 167 r 1, 335, 88 r 1, 225 r 2
B 95 r 1, 66, 213 r 2; 419 r 1, 255,
232 r 1
C 117 r 2, 324 r 1, 108 r 3, 381 r 1, 144 r 3,
93 r 2

Page 16 Division
A 3421, 2112, 2313, 4123, 1321;
418, 311, 642, 311, 421;
912 r 1, 911 r 1, 722 r 2, 832, 611 r 1;
1311 r 2, 1712, 1711 r 1, 3943, 2722 r 1;
1731 r 2, 3781 r 1, 2752 r 2, 1261 r 3,
2391 r 3;
1636 r 1, 262, 1754 r 2, 1569, 1673 r 1;
1094 r 3, 1088, 1093, 3083 r 2, 1097 r 1;
1607 r 3, 1908, 2507, 906, 1307 r 3;

1640 r 3, 4880 r 1, 1470 r 3, 1950 r 2,
1570 r 4;
970 r 2, 909, 2081 r 1, 2068 r 1, 908 r 3;
4017 r 1, 1801 r 1, 3016, 1401, 701;
2007 r 1, 1005, 2009, 1009, 2008 r 2
B 623 r 1, 1267 r 1; 1007 r 2, 2080 r 1
C 1002, 681, 1500

Page 17 Division by 7
A 2, 7, 3, 9, 0; 5, 8, 1, 4, 6
B 6, 6, 6, 5, 5; 6, 4, 3, 2, 6;
4, 5, 1, 5, 5; 4, 4, 2, 6, 3
C 7 r 6, 7 r 4, 1 r 5, 8 r 5, 4 r 4;
7 r 3, 4 r 2, 8 r 6, 2 r 6, 4 r 5;
1 r 4, 8 r 4, 7 r 2, 1 r 6, 5 r 5;
7 r 5, 4 r 6, 7 r 1, 5 r 6, 4 r 3
D 13 r 4, 12 r 3, 11 r 2, 11 r 6, 13 r 1,
11 r 1, 12 r 5; 19 r 2, 20 r 6, 23 r 3,
21 r 4, 26, 24 r 5; 134 r 5, 58 r 2, 81,
97 r 1, 126 r 2, 51 r 5; 220 r 3, 248 r 3,
280 r 2, 258 r 2, 235 r 5; 900 r 6,
490 r 5, 627, 846 r 6, 395 r 4;
1229 r 4, 1050, 1263 r 5, 1347, 1386 r 5
E 610 r 1, 129 r 6, 601 r 3, 205, 127 r 3

Page 18 Division by 8
A 3, 9, 0, 8, 2; 6, 1, 7, 4, 5
B 7, 7, 6, 7, 7; 7, 5, 5, 3, 7;
6, 3, 5, 6, 5; 6, 4, 6, 4, 4; 2, 4, 6, 2
C 1 r 7, 7 r 7, 6 r 4, 2 r 4, 8 r 7, 3 r 6;
1 r 4, 6 r 6, 7 r 4, 1 r 3, 6 r 5, 8 r 6;
2 r 7, 1 r 2, 6 r 7, 7 r 6, 2 r 6, 6 r 2;
6 r 3, 1 r 6, 7 r 5, 2 r 5, 1 r 5, 3 r 7
D 11 r 7, 10 r 7, 12, 11, 11 r 5, 12 r 2,
11 r 1; 18 r 3, 23 r 2, 24, 17 r 6, 13 r 2,
19 r 3; 118 r 2, 121 r 4, 100 r 6, 113 r 5,
124, 119 r 6; 245 r 2, 147 r 3, 236 r 6, 219
r 6, 204; 938 r 2, 791 r 2, 718 r 5, 884,
594; 1112 r 7, 1155 r 3, 1133 r 7, 1093 r 5,
1209 r 1
E 540, 1130 r 1, 91 r 1, 211 r 2

Page 19 Division by 9
A 54, 81, 18, 36, 63; 0, 45, 9, 27, 72
B 7 r 7, 2 r 8, 5 r 8, 3 r 6, 2 r 5, 4 r 6,
2 r 2; 3 r 3, 1 r 3, 6 r 8, 4 r 7, 8 r 8,
5 r 5, 6 r 6; 1 r 4, 3 r 7, 2 r 3, 1 r 2,
3 r 5, 2 r 4, 1 r 7; 5 r 7, 6 r 7, 5 r 6,
2 r 6, 7 r 8, 4 r 5, 4 r 4; 1 r 5, 2 r 7,
4 r 8, 3 r 8, 1 r 6, 1 r 8, 3 r 4
C 11 r 7, 13 r 6, 16, 21 r 3, 15 r 2,
20 r 7; 19 r 4, 16 r 2, 20 r 1, 12 r 2,
22 r 1, 18 r 6; 68 r 7, 91 r 1, 36 r 8,
29 r 3, 75 r 1, 54 r 2; 62 r 5, 44 r 5,
58 r 7, 28 r 3, 85 r 2, 34 r 3; 144 r 5,
141 r 5, 129 r 4, 194 r 4, 182 r 1;
910 r 3, 120 r 3, 157 r 3, 850 r 7, 205 r 3;
671, 255 r 5, 438 r 3, 607 r 3, 730 r 8
D 41 r 7, 467 r 4, 438 r 1, 75 r 5;
586 r 5, 89 r 8, 108 r 4, 920;
52, 238, 211 77, 102

Page 20 Division by 11
A 3, 6, 8, 1, 9; 7, 5, 0, 4, 2
B 9, 9, 7, 8, 9; 4, 7, 8, 6, 9; 8, 7, 8, 9, 7;
5, 9, 4, 8, 7; 5, 9, 6, 6, 7; 3, 3, 4, 9, 6;
6, 5, 5, 8, 8
C 13 r 7, 12 r 9, 15 r 8, 17 r 5, 15,
16 r 10; 61 r 3, 53 r 7, 79, 43 r 9,
30 r 4, 68 r 4;
9 r 4, 9 r 6, 9 r 10, 9 r 9, 9 r 8, 9 r 7;
146 r 1, 180 r 3, 157 r 9, 170, 138 r 7;
659, 200 r 7, 738 r 7, 903 r 3, 410;
97 r 7, 93 r 6, 93 r 7, 96, 96 r 7
D 312, 55, 531, 32; 570, 834, 28, 82;
572 r 2, 85 r 7, 200 r 1, 74, 99 r 9

Page 21 Division by 12
A 9, 2, 3, 6, 1; 5, 7, 8, 0, 4
B 3 r 7, 4 r 8, 2 r 9, 8 r 8, 6 r 9, 8 r 5;
7 r 6, 3 r 6, 6 r 10, 8 r 6, 4 r 4, 8 r 9;
1 r 8, 7 r 7, 4 r 3, 8 r 4, 6 r 9, 8 r 7;
4 r 7, 3 r 5, 1 r 9, 4 r 2, 4 r 9, 4 r 10;
7 r 9, 7 r 10, 4 r 5, 3 r 10, 7 r 8, 2 r 6;
2 r 8, 3 r 9, 4 r 6, 3 r 4, 6 r 8, 3 r 8
C 62 r 5, 57, 31 r 7, 30 r 4, 70 r 11,
77 r 5; 39 r 10, 50 r 7, 80 r 3, 60 r 9,
18, 34 r 6; 9 r 5, 9 r 3, 9 r 4, 9 r 9, 9 r 2,
9 r 7; 170 r 6, 191 r 9, 182 r 8, 184 r 2,
161 r 5; 808 r 8, 627 r 6, 563 r 4, 243 r 5,
398 r 9; 97 r 1, 85 r 4, 89 r 4, 99 r 5,
95 r 10
D 601 r 4, 283 r 10, 67 r 4; 600 r 6, 350,
736 r 1; 92 r 7, 33 r 8
E 173, 82, 355, 561, 33

Page 22 Division
A 46 r 1, 8 r 1, 15 r 3, 28 r 1, 12 r 6, 9;
7 r 5, 9 r 3, 8 r 1, 8 r 10, 7 r 4, 7 r 8;
365, 248 r 1, 137 r 3, 142 r 2, 229 r 2, 174;
39, 27, 54 r 1, 39 r 2, 68 r 1, 29 r 2;
69 r 9, 93 r 3, 39 r 10, 68 r 2, 83 r 5,
63 r 9; 1494 r 2, 1971 r 1, 2672 r 1,
1913 r 2, 1375 r 4; 151 r 8, 162 r 10,
193 r 6, 126 r 5, 242 r 7; 963 r 2,

907 r 1, 915 r 5, 1572 r 3, 931 r 2;
307, 470, 906, 830, 700;
1007, 1304, 1230, 1052, 3006
B 29, 669 r 1, 580, 250,
0 (nothing to add)
C 525 r 6, 94 r 3, 133, 181 r 4, 673 r 4,
85 r 5, 601 r 1, 69 r 6

Page 23 Notation
A 7 thousands, 7 hundreds, 7 units,
7 tens, 7 units; 7 hundreds, 7 tens,
7 thousands, 7 hundreds, 7 thousands
B 6541, 8310, 9743, 9642, 3100;
7710, 9742, 8310, 6532, 7642
C three thousand six hundred and twenty;
seven hundred and fifty; ninety;
eight hundred and ten; four thousand
two hundred and seventy; eight thousand
and eighty;
five hundred; four thousand
six hundred; five thousand; eight hundred;
nine thousand six hundred; seven
thousand two hundred; seven hundred;
one thousand two hundred
D thirty-seven; four; four hundred and
twenty-seven; six hundred and thirty-two;
fifty-eight; seven;
thirty-four; three; seventy-one;
eight; fifty-two
E multiply by 10, divide by 10;
divide by 100, multiply by 10;
multiply by 100, divide by 10
F 55, 450, 13; 102, 63, 145;
4100, 600, 920; 3200, 1040, 1600

Page 24 Number series and equations
A 4, 8, **12**, 16, 20, **24**, 28, **32**;
7, 14, 21, 28, **35**, **42**, **49**;
9, 12, **15**, **18**, 21, 24, **27**;
48, **60**, 72, 84, 96, **108**, **120**;
81, 72, 63, **54**, **45**, 36, **27**;
88, 80, 72, **64**, **56**, **48**;
50, **60**, 70, **80**, 90, **100**;
12, 14, 16, **18**, 20, 22, **24**;
24, 22, 20, **18**, **16**, **14**;
100, 90, **80**, 70, **60**, 50, **40**;
5, 10, 15, **20**, **25**, **30**, 35;
32, **28**, 24, **20**, 16, 12, **8**;
66, **55**, 44, 33, **22**, **11**;
55, 66, 77, **88**, **99**, **110**;
24, 32, **40**, **48**, 56, 64, **72**;
18, 27, 36, **45**, **54**, 63, 72;
48, 42, **36**, 30, **24**, 18, **12**;
77, **70**, 63, **56**, 49, **42**, 35
B 3, 6, 10, 13, 17, **20**, **24**, **27**, **31**;
52, 47, 39, 34, 26, **21**, **13**, **8**, **0**;
2, 8, 10, 16, 18, **24**, **26**, **32**, **34**;
62, 61, 55, 54, 48, **47**, **41**, **40**, **34**
4, 11, 19, 26, 34, **41**, **49**, **56**, **64**;
24, 22, 19, 17, 14, **12**, **9**, **7**, **4**
8, 13, 22, 27, 36, **41**, **50**, **55**, **64**;
54, 50, 43, 39, 32, **28**, **21**, **17**, **10**
C 3, 2, 8; 6, 6, 12; 5, 4, 4; 3, 6, 2;
4, 6, 1; 6, 10, 5; 5, 2, 12; 2, 4, 4;
1, 6, 4

Page 25 Fractions
A 1a $\frac{1}{4}$ b $\frac{3}{4}$ 2a $\frac{1}{2}$ b $\frac{1}{2}$ 3a $\frac{3}{4}$ b $\frac{1}{4}$ 4a $\frac{1}{4}$ b $\frac{3}{4}$
B D, B; D, C
C 3 cm, $4\frac{1}{2}$ cm, 3 cm; $1\frac{1}{2}$ cm, $1\frac{1}{2}$ cm,
9 cm
D 3p, 2, 9p, 10; 4, 12p, 5, 18;
20p, 10, 30, 9
E 1, $\frac{1}{2}$, $\frac{3}{4}$, 1; $\frac{3}{4}$, $\frac{1}{2}$, $\frac{1}{2}$, $\frac{1}{4}$; $1\frac{1}{2}$, $1\frac{1}{4}$, $\frac{3}{4}$, $\frac{3}{4}$

Page 26 Fractions
A **1a** $\frac{1}{8}$ b $\frac{7}{8}$ **2a** $\frac{5}{8}$ **b** $\frac{3}{8}$ **3a** $\frac{7}{8}$ **b** $\frac{1}{8}$
4a $\frac{3}{8}$ b $\frac{5}{8}$
B 8, 24, 16, 32
C 13, 19, 9, 21
D 2, 6, 4, 4
E 10, 14, 20, 12
F 10, $\frac{5}{8}$, 6, $\frac{1}{4}$
G 6, 21, 4, 12; 18, 2, 15, 3; 8, 6, 14, 12

Page 27 Fractions
A F, F, A, F, G; C, G, G, H, C
B 10 cm, 9 cm, 3 cm, 1 cm; $2\frac{1}{2}$ cm, 3 cm,
1 cm, 12 cm; 6 cm, 2 cm, 6 cm, 3 cm;
7 cm, 7 cm, 5 cm, 1 cm
C 24p, 42, 15 g, 9; 15 cm, $11\frac{1}{2}$ cm, 8 cm,
$5\frac{1}{2}$; 36, 12p, 77, 60 cm;
7p, 12, 15 g, 70 g

Page 28 Fractions
A **1** $\frac{1}{3}$ **2** $\frac{1}{6}$ **3** $\frac{2}{3}$ **4** $\frac{5}{6}$ **5** $\frac{2}{3}$ **6** $\frac{5}{6}$
B 2, 6, 4; 3, 1, 2; 6, 3
C **1** 2, 1 **2** 4, 2 **3** 4, 2
D 4, 6, 2; 10, 8, 12

Page 29 Fractions
A 2, 2, 8, 2, 6; 6, 2, 3, 3, 4; 4, 3, 4, 4, 4
B $\frac{1}{8}$, 4; 28, 8, 20; 16, 12
C 30; 15, 25, 10, 20

Page 30 Fractions
A 1a $\frac{2}{5}$ **b** $\frac{3}{5}$ **2a** $\frac{1}{10}$ **b** $\frac{9}{10}$ **3a** $\frac{8}{10}$
b $\frac{2}{10}$ **4a** $\frac{7}{10}$ **b** $\frac{3}{10}$ **5a** $\frac{3}{5}$ **b** $\frac{2}{5}$
6a $\frac{1}{5}$ **b** $\frac{4}{5}$ **7a** $\frac{3}{10}$ **b** $\frac{7}{10}$ **8a** $\frac{9}{10}$ **b** $\frac{1}{10}$ **9a** $\frac{6}{10}$ **b** $\frac{4}{10}$
B 10, 2, 2, 6, 4; 3, 8, 1, 5, 4; 7, 3, 3, 2; 4, 1, 2, 9
C 5, 63 cm, 4 g, 45; 18p, 28, 18, 84; 18 g, 12p, 72, 27

Page 31 Fractions
A **1** $\frac{1}{2}$ **2** $\frac{1}{3}$ **3** $\frac{1}{6}$ **4a** 15 **b** 10 **c** 5
B **1** $\frac{3}{5}$ **2** $\frac{2}{5}$ **3** 24
C **1** $\frac{1}{4}$ 2 $\frac{3}{4}$ **3a** 15p **b** 45p
D **1a** $\frac{3}{10}$ **b** $\frac{2}{10}$ ($\frac{1}{5}$) **c** $\frac{5}{10}$ ($\frac{1}{2}$)
2a 12 **b** 8 **c** 20

Page 32 Decimal notation
A **1** 0·2, 0·5, 0·7, 0·3, 0·6; 0·9, 0·1, 0·4, 0·8
2 0·6, 0·4, 0·3; 0·9, 0·8, 0·5; 0·2, 0·7
3 4·7, 8·3, 2·5, 7·9, 3·4; 1·6, 8·2, 5·8, 6·1
4 8·5, 9·1, 6·2; 3·9, 5·3, 7·4; 9·8, 4·6, 2·7
5 0·15, 0·23, 0·67, 0·21;
0·19, 0·83, 0·52, 0·32
6 0·06, 0·03, 0·08, 0·09;
0·04, 0·07, 0·05, 0·02; 0·01
7 0·11, 0·01, 0·43, 0·29;
0·07, 0·03, 0·09, 0·02;
0·91, 0·88
8 0·19, 0·06; 0·42, 0·09
9 4·07, 9·15, 12·37, 8·09; 11·71, 4·03
10 4·06, 19·15, 52·03, 8·16
B six tenths, six hundredths, nine units, two hundreds; six hundredths, three tens, two tenths, six units; one hundred, two tenths, nine hundredths, four tens

Page 33 Decimal notation
A 0·4, 7·1, 1·01; 4, 30·2, 18·7; 0·11, 14·2
B 11·10, 11·05, 1·11, 1·10, 1·01, 0·11;
2·01, 1·22, 1·2, 1·12, 1·02, 1·01;
16·32, 16·3, 16·23, 16·2, 16·09, 16·03;
101·99, 101·9, 101·09, 100·99, 100·9, 100·09
C 423, 420·3, 762·7, 858·6, 1432·7
D 2603, 4263, 1805, 20, 57
E 2·46, 9·1, 30·31, 4·2, 27·16
F 32·6, 4·21, 2·02, 34, 20·1
G $\div$ 10, $\div$ 100, $\times$ 10, $\div$ 100, $\div$ 10
H $\times$ 100, $\div$ 10, $\div$ 10, $\div$ 100, $\times$ 10

Page 34 Decimals
A 20·52, 144·17; 110·17, 13·06;
116·38, 113·66; 235·72, 23·73
B 5·14, 13·73; 8·78, 98·24;
50·34, 5·37; 98·93, 11·05
C 15·8, 1·57; 0·61, 13·63;
5·57, 11·03; 9·13, 104·22
D 24·28, 71·2; 22·83, 34·02;
53·2, 37·31; 43·68, 60·01
E 2·13, 99·13; 5·82, 0·91;
6·67, 8·11; 3·74, 25·93
F 5·7, 33·04, 103·5

Page 35 Money – composition to £1·00
A 17p – **2p, 1p** 51p – **5p, 2p, 2p**;
52p – **5p, 2p, 1p** 28p – **2p**;
27p – **2p, 1p** 69p – **1p**; 45p – **1p** 82p – **5p, 2p, 1p**; 21p – **2p, 2p** 21p – **2p, 2p**;
62p – **5p, 2p, 1p** 33p – **1p**; 31p – **5p, 2p, 2p**;
22p – **2p, 1p**; 41p – **2p, 2p**; 93p – **1p**
B 95p, 99p; 87p, 80p; 98p, 97p

Page 36 Money – composition to £1·00
A 88p, 76p, 54p, 60p, 47p
B 18p – **10p, 5p, 2p, 1p**
15p – **10p, 2p, 2p, 1p**
22p – **10p, 10p, 1p, 1p** or **10p, 5p, 5p, 2p**
30p – **10p, 10p, 5p, 5p**
16p – **5p, 5p, 5p, 1p** or **10p, 2p, 2p, 2p**
5p – **2p, 1p, 1p, 1p**
57p – **50p, 5p, 1p, 1p**
75p – **50p, 10p, 10p, 5p**;
8p – **5p, 1p, 1p, 1p** or **2p, 2p, 2p, 2p**
35p – **10p, 10p, 10p, 5p** or **20p, 5p, 5p, 5p**
80p – **50p, 10p, 10p, 10p** or **20p, 20p, 20p, 20p**
64p – **50p, 10p, 2p, 2p**
19p – **10p, 5p, 2p, 2p**
32p – **10p, 10p, 10p, 2p** or **20p, 5p, 5p, 2p**
12p – **5p, 5p, 1p, 1p**
9p – **5p, 2p, 1p, 1p**
C 43p – **5p, 2p** 37p – **10p, 2p, 1p**
29p – **20p, 1p** 46p – **2p, 2p**
22p – **20p, 5p, 2p, 1p** 28p – **20p, 2p**;
33p – **10p, 5p, 2p** 10p – **20p, 20p**
38p – **10p, 2p** 27p – **20p, 2p, 1p**
11p – **20p, 10p, 5p, 2p, 2p** 35p – **10p, 5p**

D 76p – **20p, 2p, 2p**
42p – **50p, 5p, 2p, 1p** 35p – **50p, 10p, 5p**
67p – **20p, 10p, 2p, 1p** 24p – **50p, 20p, 5p, 1p** 14p – **50p, 20p, 10p, 5p, 1p**
87p – **10p, 2p, 1p** 93p – **5p, 2p**
E Check your child's answers add up to 50p and £1·00.

Page 37 Money – addition and subtraction
A 35p, 53p, 44p, 51p, 64p, 82p, 73p;
61p, 70p, 51p, 81p, 94p, 91p, 71p;
81p, 93p, 86p, 87p, 80p, 96p, 81p;
57p, 78p, 83p, 90p, 86p, 89p, 88p
B 22p, 33p, 60p, 71p, 82p, 68p, 84p;
28p, 76p, 51p, 77p, 49p, 58p, 70p;
18p, 41p, 33p, 33p, 41p, 5p, 25p;
19p, 26p, 6p, 31p, 49p, 23p, 58p
C 98p, 15p, 23p, 98p, 58p, 36p

Page 38 Money – multiplication and division
A 14p, 24p, 54p; 80p, 36p, 56p;
55p, 27p, 20p; 54p, 88p, 84p;
27p, 96p; 28p, 80p, 63p; 36p, 24p, 40p;
33p, 15p, 24p; 77p, 36p, 36p;
70p, 88p, 57p, 90p, 91p, 92p, 51p;
34p, 84p, 76p, 69p, 64p, 68p
B 24p, 9p, 27p, 7p, 14p, 38p, 19p;
21p, 19p, 12p, 13p, 42p, 12p, 18p;
16p, 16p, 12p, 11p, 13p, 27p, 13p
C **1** 72p **2** 23p **3** 72p **4** 14p **5a** 52p
B 78p **c** 39p **d** 9 1p **e** 26p **6** 18p each

Page 39 Pounds and pence
A 2·00, 5·00, 7·00, 6·00;
9·00, 3·00, 8·00, 4·00
B 1·04, 3·05, 1·43; 2·36, 6·24, 3·79;
4·76, 2·17, 9. 14; 8·07, 7·63, 2. 07
C £3·45 = £**3** + **45**p, £6·17 = £**6** + **17**p;
£7·06 = £**7** + **6**p, £9·73 = £**9** + **73**p;
£12·14 = **£12** + **14p**, £15·48 = £**15** + **48**p;
£2·72 = £**2** + **72**p, £8· 05 = £**8** + **5**p
D £4·17 = **417**p, £8·12 = **812**p,
£9·08 = **908**p; £8·36 = **836**p,
£6·18 = **618**p, £3·12 = **312**p;
£2·04 = **204**p, £5·42 = **542**p,
£5·23 = **523**p; £6·25 = **625**p,
£4·38 = **438**p, £4·77 = **477**p
E 2+7+2, 6+4+2, 9+0+4, 7+7+2, 8+9+6
F 270, 436, 753, 947, 1036;
92, 34, 25, 67, 83

Page 40 Pounds and pence
A 17p, 27p, 36p, 76p, 89p, 5p
B £0·26, £0·07, £0·34, £0·08, £0·52, £0·05
C £4·27, £0·76; £1·18, £0·06;
£2·04, £0·96
D seven pounds forty-five, seventy-two pence, fourteen pounds four pence, three pounds sixty-two; twelve pence, five pounds thirty-two, nine pounds twenty-six, seven pence
E **a** £4·36, 414p, £4·00, 47p, £0·44;
b £5·50, £5·27, 505p, £0·61, 59p;
c £22·00, £2·20, £2·02, 200p, 22p
F £1·09, £1·21, £1·16, £0·92, £1·29, £1·10;
£0·70, £1·21, £1·15, £1·03, £0·99, £1·16

Page 41 Addition and subtraction – £ and p
A £0·60, £0·85, £0·89, £0·82, £0·88;
£1·06, £1·26, £1·33, £1·30, £1·23;
£6·73, £9·00, £9·39, £9·99, £13·03;
£8·46, £7·04, £10·63, £10·21, £18·02
B £0·05, £0·13, £0·28, £0·17, £0·29;
£1·16, £0·60, £0·86, £0·63, £0·57;
£2·08, £3·51, £3·00, £0·79, £1·03
C £7·05, £2·78, £16·65, £6·08, £7·24, £15·55

Page 42 Multiplication – £ and p
A £0·90, £0·96, £0·96, £0·94, £0·90;
£0·95, £0·84, £0·91, £0·81, £0·88;
£4·24, £4·56, £3·80, £3·12, £5·76;
£2·88, £3·35, £6·51, £2·86, £3·84;
£17·16, £27·60, £24·56, £38·88, £9·54;
£6·21, £7·44, £22·26, £12·86, £44·91
B £12·66, £23·38; £19·02, £109·44;
£19·62, £101·16
C £25·38, £17·60
D £96·48, £27·78
E £8·13, £0·81

Page 43 Division – £ and p
A £0·06, £0·24, £0·14, £0·17, £0·29;
£0·07, £0·12, £0·10, £0·46, £0·06;
£0·12, £0·13, £0·09, £0·07, £0·26;
£0·16, £0·23, £0·47, £0·12, £0·22;
£0·26, £0·17, £0·15, £0·18, £0·32;
£0·44, £0·17, £0·38, £0·22, £0·25;
£4·36, £1·81, £1·39, £0·61, £1·20;
£2·70, £3·17, £5·62, £3·93, £2·19;
£4·90, £3·83, £2·87, £3·16, £2·81

B £0·96, £0·68; £11·03, £9·41; £0·39, 4·02; £0·76, £3·01; £6·70, £6·04
C £4·72, £2·32; £0·92, £0·34; £12·03, £10·71

Page 44 Buying
1 £1·40 **2** £7·16 **3a** 28p **b** 25p **c** 55p **4** £4·55 **5** £10·50 **6a** £5·00 **b** £6·66 **c** £2·88 **d** £3·25 **e** £2·10 **7** £1·77 **8** £2·58 **9** 6 kites, 10p change **10** 3 aeroplanes, £1·66

Page 45 Capacity
A 1 1000 ml, 2000 ml, 4000 ml, 6000 ml; 9000 ml, 7000 ml, 3000 ml, 5000 ml; 500 ml, 250 ml, 1500 ml, 1250 ml; 750 ml, 2250 ml, 2750 ml, 1750 ml
2 500 ml, 400 ml, 900 ml, 700 ml; 300 ml, 800 ml, 200 ml, 100 ml; 2400 ml, 1600 ml, 3800 ml, 4500 ml; 6100 ml, 5300 ml, 8800 ml, 2700 ml
B Check your child's answers all add up to 1 litre.
C Check your child's answers all add up to 500 ml.
D 1 medicine and small milk **2** 4 bottles **3** lemonade and pop **4** 125 ml **5** 3 **6** small milk, pop and cordial

Page 46 Length
A 200 cm, 900 cm, 600 cm, 400 cm, 700 cm 1200 cm, 1600 cm
B 4 m, 3 m, 5 m, 1 m, 8 m, 13 m
C 50, 25, 75; 250, 125, 175; 450, 225, 275; 650, 425, 775
D $\frac{1}{2}$, $\frac{1}{4}$, $\frac{3}{4}$; $3\frac{1}{2}$, $5\frac{1}{4}$, $3\frac{3}{4}$; $1\frac{1}{2}$, $3\frac{1}{4}$, $4\frac{3}{4}$
E 123, 216; 405, 308; 936, 829
F 2 m 25 cm, 5 m 18 cm; 3 m 12 cm, 6 m 54 cm; 8 m 6 cm, 9 m 31 cm
G 5·32 m, 1·96 m, 0·92 m; 4·02 m, 0·52 m, 3·07 m; 3·9 m, 0·07 m, 14·26 m; 426 cm, 816 cm, 1027 cm; 1206 cm, 9 cm, 341 cm; 27 cm, 77 cm, 101 cm
H 0·5 m, 0·25 m, 0·75 m, 0·01 m, 0·23 m

Page 47 Length
A 5 cm, 4 cm, 6 cm; 3 cm, 9 cm, 2 cm; 13 cm, 22 cm, 19 cm
B 70 mm, 80 mm, 90 mm; 60 mm, 50 mm, 30 mm; 150 mm, 3200 mm, 5300 mm
C 53 mm = 5 cm 3 mm = 5·3 cm, 7 mm = 0 cm 7 mm = 0·7 cm; 86 mm = 8 cm 6 mm = 8·6 cm, 4 mm = 0 cm 4 mm = 0·4 cm; 131 mm = 13 cm 1 mm = 13·1 cm, 6 mm = 0 cm 6 mm = 0·6 cm; 268 mm = 26 cm 8 mm = 26·8 cm 320 mm = 32 cm 0 mm = 32 cm
D 41 mm, 37 mm, 9 mm, 8 mm 142 mm, 56 mm
E 7 m, 9 m, 3 m; 4·5 m, 2·5 m, 8·5 m; 5·25 m, 8·25 m, 1·25 m; 9·75 m, 3·75 m, 4·75 m
F 6500 mm, 5500 mm, 3500 mm, 2750 mm, 1750 mm, 4250 mm, 7250 mm, 9250 mm
G 6670 mm = 6 m 670 mm, 5026 mm = 5 m 26 mm, 2340 mm = 2 m 340 mm; 8026 mm = 8 m 26 mm, 2726 mm = 2 m 726 mm, 8791 mm = 8 m 791 mm
H 4 m 135 mm = 4135 mm, 6 m 296 mm = 6296 mm, 8 m 196 mm = 8196 mm; 16 m 176 mm = 16 176 mm, 5 m 17 mm = 5017 mm, 9 m 5 mm = 9005 mm

Page 48 The kilometre
A 1275 m, 3536 m, 4326 m; 8824 m, 2056 m, 7036 m; 5004 m, 6008 m, 4500 m, 3750 m; 2750 m, 9500 m
B 3 km 760 m, 2 km 430 m, 1 km 641 m; 8 km 252 m, 5 km 862 m, 4 km 26 m; 7 km 54 m, 1 km 6 m, 3 km 2 m, 6 km 8 m

Perimeter
C A a 7·5 cm **b** 2·5 cm **c** 20 cm, **B a** 4·5 cm **b** 4·5 cm **c** 18 cm, **C a** 3·5 cm **b** 2 cm **c** 11 cm, **D a** 6 cm **b** 2·5 cm **c** 17 cm, **E a** 7 cm **b** 2·5 cm **c** 19 cm, **F a** 5·5 cm **b** 4·5 cm **c** 20 cm, **G a** 5·5 cm **b** 1·5 cm **c** 14 cm

Page 49 Area and perimeter
A A 15 cm^2, **B** 8 cm^2, **C** 9 cm^2, **D** 17 cm^2, **E** 29 cm^2, **F** 13·5 cm^2, **G** 9 cm^2
B A 24 cm^2, **B** 6 cm^2, **C** 18 cm^2, **D** 4 cm^2, **E** 16 cm^2

Page 50 Area and perimeter
A a 27 cm **b** 38 cm^2, **B a** 17 cm **b** 17·5 cm^2, **C a** 15 cm **b** 12·5 cm^2, **D a** 21 cm **b** 17 cm^2, **E a** 27 cm **b** 31·5 cm^2, **F a** 17 cm **b** 15 cm^2, **G a** 19 cm **b** 17·5 cm^2, **H a** 15 cm **b** 6·5 cm^2, **I a** 18 cm **b** 18 cm^2

Page 51 Scale measurement
A 1 20 cm **2** 4 cm **3** 11 cm **4** 8 cm **5** 16 cm **6** 9 cm
B 10 cm, 15 cm, 9 cm, 7 cm, 3 cm, 6 cm
C 30 cm, 45 cm, 65 cm, 55 cm, 70 cm
D 4·5 cm, 6 cm, 5·5 cm, 2 cm, 5 cm, 2·5 cm
E 48 km

Page 52 Scales and maps
A 1 150 km **2** 100 km; **3** 60 km **4** 30 km; **5** 50 km **6** 60 km; **7** 160 km **8** 70 km; **9** 220 km **10** 120 km; **11** 70 km **12** 60 km
B 42·5 km, 55 km, 62·5 km, 22·5 km, 65 km; 3 cm, 8 cm, 5 cm, 12 cm, 3·5 cm, 10·5 cm
C 1 cm to 1 m, 1 cm to $\frac{1}{2}$ km, 1 cm to $\frac{1}{5}$ km, 1 cm to $\frac{1}{4}$ m, 1 cm to $\frac{1}{3}$ km

Page 53 Mass
A 500 g × 10; 200 g × 25; 100 g × 50; 50 g × 100; 20 g × 250; 10 g × 500
B 2000 g, 8500 g, 6500 g, 10 000 g, 3500 g, 7000 g
C 4 kg, 6 kg, 8·5 kg, 2·5 kg, 5 kg, 3·5 kg
D 4 kg 260 g, 8 kg 120 g, 2 kg 50 g, 3 kg 160 g, 7 kg 5 g
E 2360 g, 3115 g, 5070 g, 4033 g, 6005 g,
F 3200 g, 4100 g, 6300 g, 5400 g, 8600 g,
G 64 g, 271 g, 153 g, 209 g, 468 g, 433 g,
H 1 kg 471 g, 126 g, 2 kg 21 g; 9 kg 741 g, 6 kg 6 g
I 1 kg 750 g, 1 kg 400 g, 1 kg 100 g, 5 kg 500 g, 50g

Page 54 Comparison – other measures
1 300 oranges **2** 287·5 l **3** 6 lengths: 20 cm left **4** 5·4 l **5** 3·65 kg **6** 34 packs **7** 4·5 m **8** 58 l **9** 250 bottles **10** 100 bags **11** 25 bags **12** 8·4 l **13** 1·36 m (136 cm) **14** 4 lengths **15** 1·7 kg **16** 200 steps **17** 1·050 kg **18** 60 doses **19** 1·5 kg **20a** 33 **b** 4 cm

Page 55 Shopping
la £3·20 **b** £4·80 **c** £8·00 **d** £7·20
2a 5p **b** 25p **c** 90p **d** £1·20
3a 8p **b** 48p **c** 56p **d** 72p
4a 90p **b** 40p **c** 45p **d** 75p
5a 80p **b** 96p **c** 8p **d** 40p
6a 45p **b** 30p **c** £2·55 **d** £1·65
7a £12·50 **b** £21·25 **c** £15·00 **d** £8·75
8a 30p **b** £2·70 **c** £4·20 **d** £6·00
9a £2·88 **b** £2·40 **c** £4·80 **d** £6·72
10a 48p **b** 24p **c** £1·20 **d** £1·56
11a 11p **b** 44p **c** £2·42 **d** 99p
12a £6·30 **b** £1·75 **c** £4·20 **d** £8·40
13a £1·75 **b** £1·05 **c** £2·45 **d** £3·50
14a 8p **b** £1·76 **c** £2·00 **d** £2·24
15a 5p **b** 35p **c** 45p **d** 15p
16a 28p **b** 42p **c** 63p **d** £1·12
17a 24p **b** 30p **c** 60p **d** 75p
18a 40p **b** 16p **c** 28p **d** £2·20

Page 56 Time
A 5 past 2, 20 past *7,* 12 o'clock, 25 past 4, 20 past 11; $\frac{1}{2}$ past 1, 10 past 5, 10 past 9, $\frac{1}{4}$ past 6, 25 past 10
B 25 to 10, 35 min past 9; 5 to 8, 55 min past 7; 20 to 5, 40 min past 4; 10 to 2, 50 min past 1; $\frac{1}{4}$ to 4, 45 min past 3; 5 to 12, 55 min past 11; $\frac{1}{4}$ to 2, 45 min past 1; 20 to 1, 40 min past 12; 10 to 3, 50 min past 2; 25 to 9, 35 min past 8
C Check that your child's clocks show the correct times.

Page 57 Time
A 6:10, 8:45, 3:50; 8:15, 4:40, 9:55
B 1:20, 12:05, 6:55, 2:25, 3:35
C 2:15 am, 9:35 am, 6:45 am, 4:10 am, 8:20 am; 11:50 pm, 3:30 pm, 5:40 pm, 11:25 pm, 8:55 pm
D 12:35 pm, 6:20 pm, 3:50 am, 7:45 pm, 12:25 am, 8:55 pm, 2:15 am, 11:10 pm, 3:05 pm; 5:30 am

Page 58 Time
A a 4 min **b** 8 min **c** 11 min **d** 19 min **e** 22 min **f** 28 min **g** 32 min **h** 39 min **i** 43 min **j** 49 min **k** 53 min **l** 56 min

B 12:08, 2:38, 5:24, 6:49;
10:42, 11:48, 2:03, 8:26
C 7:44 am, 4:59 pm, 1:17 pm, 10:29 am

Page 59 Adjusting times
A 2:05, 12:09, 8:01, 4:06
B 6:56, 4:59, 7:52, 6:02
C 30 min, 19 min, 2 min, 6 min
D 25 min, 15 min; 40 min, 20 min;
45 min, 5 min

Page 60 Time – problems
A **1** 1 h 40 min **2** 24 min **3** 42 min
4 1 h 5 min **5** Benford to Markham
6 Southam to Fartown
B 2 h 30 min, 30 min;
19 min, 22 min; 1 h 13 min, 43 min;
1 h 55 min, 2 h 21 min
C 90 min, 75 min, 150 min, 270 min;
105 min, 135 min, 165 min
D 6:25 pm
E 3:10
F 8 min
G **1** 75 min **2** 3:45 pm

Page 61 The calendar
1 No, as February has 28 days, not 29
2 March, June, August, November
3 April, July **4** March, June **5** June
6 20 June
7 13 January, 10 February, 10 March,
14 April, 12 May, 9 June, 14 July,
11 August, 8 September, 13 October,
10 November, 8 December
8 January – Wednesday, March – Saturday,
May – Thursday, July – Tuesday, August –
Friday, October – Wednesday, December –
Monday
9 April – Sunday, June – Friday, September
– Saturday, November – Thursday
10 Monday **11** Tuesday **12** Tuesday

Page 62 Graphs
A 35, 36, 30, 35, 26, 32
1 History **2** French **3** Geography and Maths
4 Science **5** 10 **6** 240 **7** 194 **8** 46
B 17, 34, 41, 36, 43, 21
1 98 **2** 94 **3** 192 **4** 8 **5** football **6** netball
7 64 **8** 53 **9** hockey **10** 4

Page 63 Graphs
A **1** Heaton **2** Birford **3** Wessex and Frame
4 Birford and Imford **5** 8 **6** Heaton and
Birford
B **1** 180 **2** 19 **3** class 4 **4** class 1
C

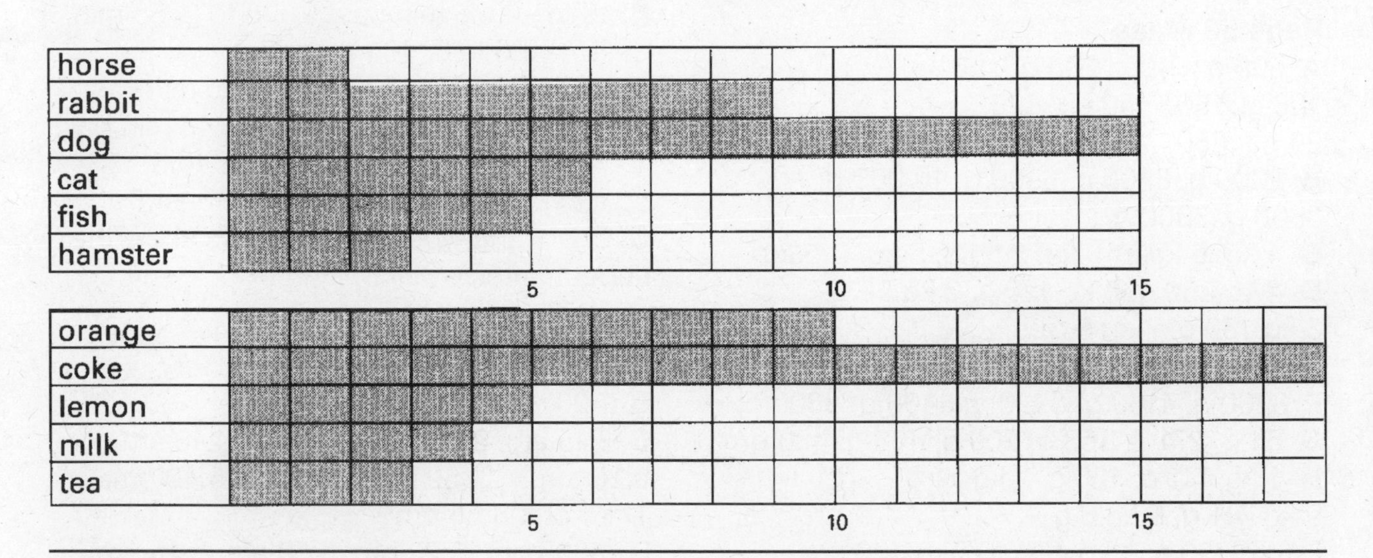

Published by Collins
An imprint of HarperCollins*Publishers* Ltd
1 London Bridge Street
London SE1 9GF

HarperCollins*Publishers*
1st Floor, Watermarque Building,
Ringsend Road, Dublin 4, Ireland

First published in 1978
This edition first published in 2012

15 14 13 12 11

ISBN 978-0-00-750549-4
The authors assert their moral right to be identified as the authors of this work.

British Library Cataloguing in Publication Data.

A catalogue record for this publication is available from the British Library.

Project managed by Katie Galloway
Production by Rebecca Evans
Page layout by Exemplarr Worldwide Ltd
Illustrated by A. Rodger
Printed in India by Multivista Global Pvt Ltd..